D1740437

MECHANICAL PROPERTIES OF
SOLIDS AND FLUIDS

MECHANICAL PROPERTIES OF SOLIDS AND FLUIDS

R. C. STANLEY, B. Sc., M. Inst. P.
Lecturer, Department of Applied Physics,
Brighton Polytechnic

LONDON   BUTTERWORTHS

THE BUTTERWORTH GROUP

ENGLAND
Butterworth & Co. (Publishers) Ltd.
London: 88 Kingsway, WC2B 6AB

AUSTRALIA
Butterworth & Co. (Australia) Ltd.
Sydney: 586 Pacific Highway, Chatswood, NSW 2067
Melbourne: 343 Little Collins Street, 3000
Brisbane: 240 Queen Street, 4000

CANADA
Butterworth & Co. (Canada) Ltd.
Toronto: 14 Curity Avenue, 374

NEW ZEALAND
Butterworth & Co. (New Zealand) Ltd.
Wellington: 26–28 Waring Taylor Street, 1

SOUTH AFRICA
Butterworth & Co. (South Africa) (Pty) Ltd.
Durban: 152–154 Gale Street

First published 1972

© Butterworth & Co. (Publishers) Ltd., 1972

ISBN 0 408 70305 9 Standard
       0 408 70306 7 Limp

Printed by photo-lithography and made in Great Britain at
the Pitman Press, Bath

PREFACE

In the past, students have had the use of several good textbooks on the general properties of matter. The modern student, however, requires a text more in keeping with his role as a technologist. He now requires a greater knowledge of the applications of the properties of matter rather than a purely academic knowledge that is divorced from much of the realities of life. Perhaps the greatest change that has occurred in this context is the increased understanding of the relation between forces and relative motions at the atomic level, and the resulting macroscopic properties of solids and fluids. It is with this aspect of the general properties of matter that this book deals. By restricting the field of study in this way, this book may more readily fit into the teaching syllabuses for both physicists and engineers.

At many colleges and universities, the subject matter of this book is covered during the first year of a degree course. Consequently a student's mathematical ability may not be very advanced. This is borne in mind in this text, where only a minimum mathematical knowledge is required with all the steps in a derivation being shown and where necessary explained. Thus HNC as well as physics and engineering degree students should find this book suitable.

As to the subject matter itself, the first three chapters deal with elasticity and plasticity. Here the concentration is mainly on the engineering metals and the development of an understanding of their properties from their structures and atomic bonding, and the nature and movement of dislocations. It is shown to what extent the motions of these dislocations can be controlled to produce a material of a required strength. Also included is the conventional study of the deformation of various bodies as related to the elastic moduli, as well as practical applications such as strain gauges and fibre composite materials. In discussing elasticity, the opportunity has been taken to introduce tensors. This has been done slowly, explaining the reasons for each step so that the student has a well-founded appreciation of their significance which will be of later use in other branches of physics. However, it is appreciated that not all students may require this refinement and therefore these chapters are presented in a way that by selective reading a student may nevertheless be able to appreciate and understand the subject matter. An alternative simplified approach is also given that avoids the use of tensors.

Similarly, in the remaining chapters, on the flow of liquids and gases, emphasis is on the more practically useful aspects and industrial methods of measurement. The large number of methods for the measurement of viscosity included in older textbooks have been largely omitted since, though of academic interest to a pure scientist, they have little or no practical importance to an applied scientist. For this reason a more detailed coverage is given of the properties of oils and their measurements, including the mechanisms of lubrication and the properties of the very important range of non-Newtonian fluids.

In the final chapter, the behaviour of gas molecules at low pressures leads to a study of the applications and techniques of both high and ultrahigh vacua. With its increasing relevance in all branches of science and technology, vacuum physics is here dealt with in a particularly practical and applied manner.

The units used throughout are SI (Système International d'Unités), although, where other units have long been in common use, values are also quoted in these alternatives. A table of conversions of older units into their new SI equivalents is also given.

Brighton   R.C.S.

CONTENTS

Chapter One

ELASTICITY

1.1 Introduction

If a force is applied to a solid so that it still remains in static equilibrium, then it will be deformed in some way. The amount of deformation will depend, obviously, on the magnitude of the applied force and how it is applied, but more basically it will depend on the structure of the solid at the atomic level. In this chapter we shall start by examining first the results that are obtained when a deforming force is applied to actual solids and secondly to what extent the results may be explained by the forces between atoms. We shall then go on to discuss more fully the meaning of stress and strain and the relationships between them. In the next chapter we shall continue by applying these ideas to specific problems.

To be able to examine and compare results means that the applied forces and the resulting deformations must be definable and reproducible in a logical manner. Let us consider a metal wire to be stretched by an axial force. The original length may be l_0 and the increase in length Δl. If now another piece of the same metal wire, but this time of length $2l_0$, is stretched by the same force as before, the resulting increase in length would be $2\Delta l$. The deformation then, in this case, is a function of the length. So that the results of tests on other samples of wires may be compared, this effect of the original length on the magnitude of the deformation must be removed. This is simply done by considering the relative deformation $\Delta l/l_0$, or relative displacement per unit distance, termed the *strain*. Similarly, the thickness of the wire will have an effect. The greater the cross-sectional area of the wire, the proportionally greater will be the force required to produce the same strain. Thus it is convenient to refer to the applied force in terms of force per unit area, in this case the cross-sectional area. The force expressed in this way is then termed the *stress*.

The sample under test need not necessarily be in the form of a wire with an applied tensile force. It could just as well be a compressive force applied axially to a rod. Nor need only axial forces be considered. The applied force could be a shear stress causing an angular deformation or shear strain to a block specimen, as shown in Figure 1.1, or it could be applied uniformly over the surface of the specimen, as for example

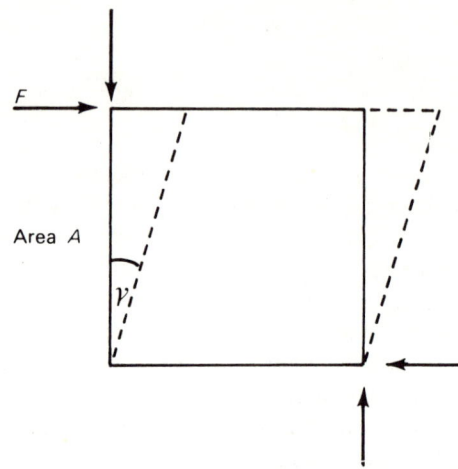

Figure 1.1.   Deformation of a cube due to a shearing force

a specimen subjected to a hydrostatic pressure, producing a volume strain. In all cases, however, there is a linear relationship between the applied stress, no matter in what manner it is applied, and the resulting strain. If the stress is doubled, so is the strain, at least to a certain magnitude of stress. The departure from linearity beyond this stress is the subject of Chapter 3. Under the conditions where stress is proportional to strain, however, the solid is said to be perfectly elastic. This linear relationship constitutes Hooke's law, published in 1678. If the stress is removed, the strain will disappear with it, but there is a certain limiting stress for the specimen under test when this will not be so. Then the solid will not return to its original dimensions, the *elastic limit* has been exceeded, and *permanent set* acquired by the specimen.

Hooke's law, that strain is proportional to the applied stress, may be applied to a wire or a rod under a tensile stress and expressed as

$$\epsilon = s\sigma \qquad (1.1)$$

where ϵ is the linear strain $\Delta l/l_0$, σ is the corresponding normal stress,

2

and *s* is a constant of proportionality called the *elastic compliance constant* or *compliance*.

Also, writing

$$c = \frac{1}{s} = \frac{\sigma}{\epsilon} \qquad (1.2)$$

gives *c* as the *elastic stiffness constant* or *stiffness* or *elastic constant*.

The elastic stiffness for a wire under a tensile stress is also termed *Young's modulus*, defined as follows:

$$\text{Young's modulus, } E = \frac{\text{normal stress, } \sigma}{\text{linear strain, } \epsilon}$$

$$= \frac{\text{applied force per unit area of cross-section}}{\text{change in length per unit length, i.e. } \Delta l/l_0} \qquad (1.3)$$

Similarly, the *bulk modulus* is defined as

$$\text{bulk modulus, } K = \frac{\text{bulk stress}}{\text{volume or bulk strain, } \theta}$$

$$= \frac{\text{pressure, } p}{\text{change in volume per unit volume, i.e. } \Delta V/V_0} \qquad (1.4)$$

Since the volume decreases as the pressure increases, and the bulk modulus is a positive quantity, a convention is sometimes used in which a negative sign is introduced into the definition. The reciprocal of the bulk modulus is termed the *compressibility*.

When applied to liquids and gases, it is often more convenient to write the expression for the bulk modulus in a different form. For example, a small change δp in pressure will produce a small change δV in a volume V of gas. Then K may be written as

$$K = \frac{\delta p}{\delta V/V} = V \frac{\delta p}{\delta V}$$

In the limit, for small changes,

$$K = V \frac{dp}{dV}$$

The third modulus in common use is the *shear modulus*, defined as

$$\text{shear modulus, } G = \frac{\text{shear stress}}{\text{shear strain}}$$

$$= \frac{\text{tangential force per unit area, } \tau}{\tan \gamma} \qquad (1.5)$$

3

This is illustrated in Figure 1.1, where a cube of face area A is subjected to a shear force F and deformed through an angle γ. In other words, shear strain is the difference in displacement of two parallel faces of the cube, divided by the perpendicular distance between them. For small shear strains, tan γ is approximately equal to the angle γ. The shear modulus is also called the *modulus of rigidity* or *torsion modulus*.

Closely related to Young's modulus is *Poisson's ratio*. If an experiment is performed to measure Young's modulus for a rod, that is, a tensile force is applied and the change in length measured, it will be found that there will be an accompanying decrease in the transverse dimensions with increasing longitudinal dimensions. The converse will also apply with an applied compressive force instead of a tensile one. The relative dimensional changes can be expressed as

$$\text{Poisson's ratio, } \nu = -\frac{\text{lateral strain}}{\text{longitudinal strain}}$$

$$= -\frac{\text{change in width per unit width}}{\text{change in length per unit length}} \qquad (1.6)$$

Since the lateral and longitudinal strains will be of opposite sign, a positive value of the ratio will result.

Other terms commonly occurring are the specific modulus and the specific strength. The *specific modulus* is equal to Young's modulus divided by the density ρ of the material, and the *specific strength* is the ultimate tensile strength (u.t.s.) – that is, the maximum tensile stress that the material can support – again divided by the density ρ. Both are in units of $m^2 s^{-2}$, that is (velocity)^2. It may be recalled that the velocity of sound in a medium is given by the square root of the quotient of elasticity and density.

From the definitions it will be seen that Young's modulus and shear modulus will have the units of stress, that is, force per unit area. The shear modulus for most materials is about one-third to one-half of Young's modulus. The bulk modulus has the same units as pressure, whilst compressibility has units of reciprocal pressure. Poisson's ratio is dimensionless and has values between 0 and $\frac{1}{2}$ for isotropic materials. Typical values for these quantities are given in Table 1.1, and a discussion of methods of measurement is given in the next chapter.

The relationship between stress and strain for a material is conveniently shown as a stress–strain curve.

Table 1.1. TYPICAL VALUES OF ELASTICITIES FOR VARIOUS MATERIALS
(PRECISE VALUES WILL DEPEND ON COMPOSITION, HEAT TREATMENT, AND METHOD OF FORMING)

Material (at 20°C)	Young's modulus, E (10^{10} N m^{-2})	Shear modulus, G (10^{10} N m^{-2})	Bulk modulus, K (10^{10} N m^{-2})	Poisson's ratio, ν	Ultimate tensile strength (10^{8} N m^{-2})	Specific modulus, E/ρ (10^{6} m^{2} s^{-2})	Specific strength, u.t.s./ρ (10^{4} m^{2} s^{-2})
Aluminium							
annealed	6·9 ⎫				0·5	25	1·8
die-cast	7·1 ⎬	2·6	7·6	0·34	2·5	26	9
drawn wire	7·0 ⎭				2–4·5	26	7·5–17
Brass	10·1	3·7	11·2	0·35	1·5–4	12	2–5
Copper	13·0	4·8	13·8	0·34	1·2–4·5	15	1·5–5
Iron							
soft	21	8·1	17·2	0·29	1–2	27	1·5–2·5
wrought	19	7·5	13·6	0·27	3–6	24	4–7·5
Lead	1·6	0·55	5·9	0·45	0·15–2	1·4	0·1–2
Phosphor bronze	12·0	4·3	19·1	0·38	2–10	13	2–11
Steel							
carbon	21·0–21·3 ⎫	7·7–8·4	16·5–18·0	0·27–0·30	7–23	25–27	10–30
low-alloy	19·9–21·4 ⎭						
Carbon fibre							
high-modulus	38–45	—	—	—	14–21	190–220	70–100
low-modulus	24–27	—	—	—	24–32	140–160	140–190
Glass fibre							
high-strength	7–8·5	—	—	—	17–26	25–34	70–100

1.2 Stress–Strain Curves

Stress–strain curves are usually given for rod or bar specimens which have been subjected to a tensile stress. The reason for this is simply that these tests are the easiest to perform. The specimen is usually in the form of a rod (Figure 1.2) with a constant cross-section over the

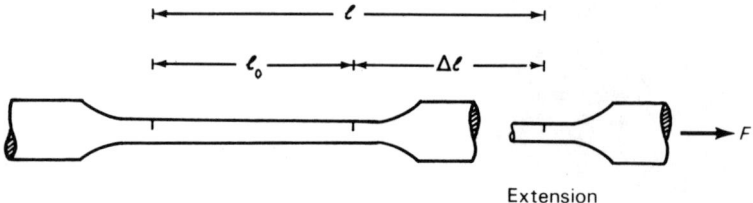

Figure 1.2. Tensile test specimen

central portion and with two marks made to designate a length l_0. A tensile load F is applied, causing the test length l_0 to extend to l. This test is continued with steadily increasing values of the load F and a continuous plot of the stress F/A_0 against the strain $\Delta l/l_0$ is made, where $\Delta l = l - l_0$. In calculating the stress, the original cross-sectional area A_0 of the test piece is normally used, even though the cross-sectional area may reduce quite drastically. For example, just before fracture, necking may occur as in Figure 1.3, where there may be a considerable localised increase in the true stress.

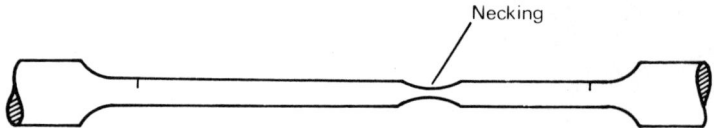

Figure 1.3. Localised failure of a tensile test specimen

If a mild steel rod is loaded in tension, a stress–strain curve similar to either Figure 1.4(a) or Figure 1.4(b) is generally observed. The increase in strain is directly proportional to the increase in stress, up to some point A, the elastic limit, but beyond this point a permanent deformation begins to appear. For stresses up to that corresponding to the point A, the specimen behaves elastically, that is, if the stress is removed, the strain will disappear almost instantaneously and there will be no permanent set. The transition to the horizontal part of the curve indicates an inelastic deformation with the metal beginning to flow plastically. (The mechanism of plastic flow constitutes the subject matter of Chapter 3, whereas in this chapter we shall be concerned with elastic behaviour of the material.) The transition varies from a gentle curve to a sharp peak, as shown. As the stress corresponding to the point A is reached, there is a sudden increase in strain owing

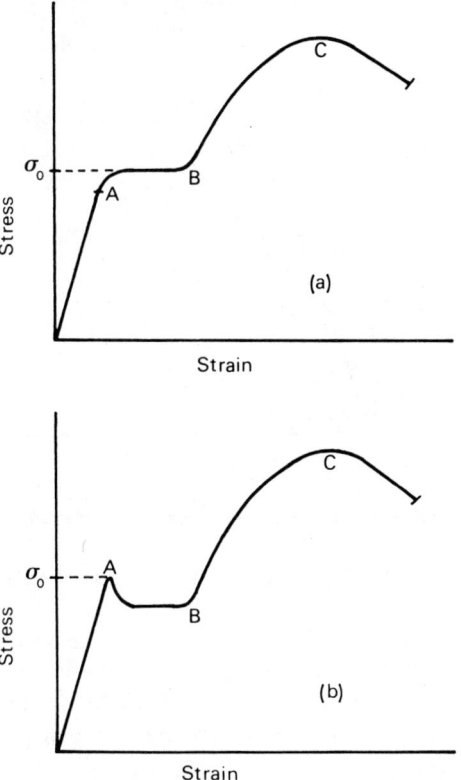

Figure 1.4.   *Characteristic shapes of the stress–strain curve for a mild steel rod specimen under tension*

to a yielding of some portion of the test piece. The onset of pronounced yielding corresponds to the *yield stress* σ_0. Beyond point B, work hardening (to be discussed in Chapter 3) increases, causing an upward turn in the curve, but at the same time the cross-sectional area is decreasing. With rapid decrease in cross-sectional area due to the onset of necking, the curve may take a downward swing before the rod finally fractures.

7

As well as yield stress, another point on the curve is of especial interest to the engineer. This is the *proof stress* or *yield strength* and is defined as the stress required to give a certain amount of plastic deformation, usually 0·1%, although values of 0·2%, 0·3%, and 0·5% are also used. The definition of the proof stress is illustrated in Figure 1.5. Proof stress has the advantage of giving a clearly defined

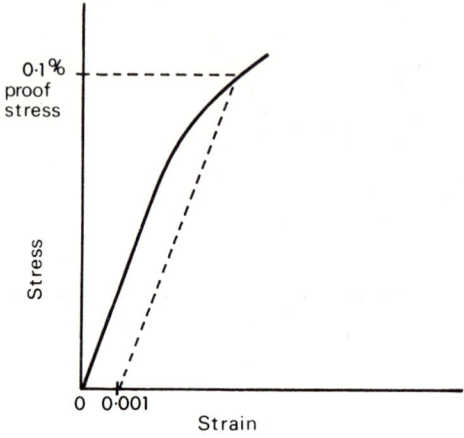

Figure 1.5.   Illustrating the definition of proof stress

point related to the elastic deformation, whereas the yield point is not always clearly discernible.

A further term used by engineers is the *ultimate tensile strength*. This is the highest value of stress occurring on the stress–strain curve and corresponds to the point C in Figure 1.4. That is, it is the maximum load that the specimen can support, divided by the original cross-section of the specimen.

For the higher-carbon steels, there is usually a gradual transition from the elastic to the plastic condition and there may not even be a recognisable yield point. In this case, it is common to refer to the end of the straight section of the curve as the *limit of proportionality*, as at B in Figure 1.6. Beyond this point, permanent set after the removal of the stress becomes appreciable.

Some materials, such as rubber, have a non-linear stress–strain relationship but are still said to be elastic in the sense that they return to their original shape when the stress is removed, as shown in Figure 1.7. Such materials show *mechanical hysteresis*, whilst the area enclosed within the curve is a measure of the energy lost in the process. This energy in a flexing tyre, for example, produces heating of the tyre. A stress will still be reached that corresponds to the elastic limit, that is, a stress that on removal does not cause the specimen to return to its

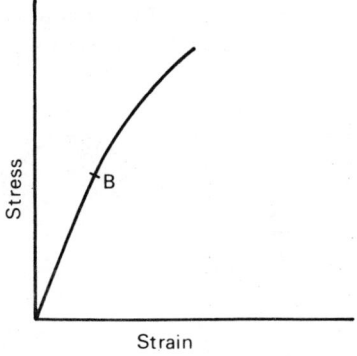

Figure 1.6.   Limit of proportionality occurring at B

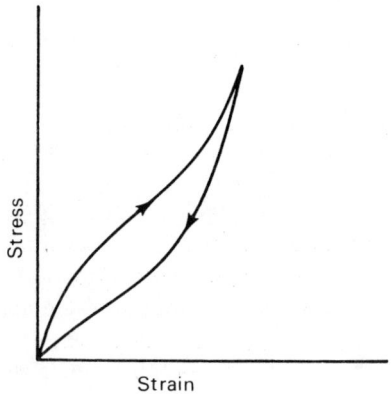

Figure 1.7.   Mechanical hysteresis in rubber

original shape. For the very soft metals, such as annealed copper or aluminium, permanent deformations occur at even the smallest stresses, as shown in Figure 1.8. Here there is no linear relationship between stress and strain and it becomes necessary to draw a tangent to the curve at the origin to determine $d\sigma/d\epsilon$ to obtain a value for Young's modulus E. The strain ϵ in the figure corresponding to a stress σ may

9

be considered to be the sum of the elastic strain $\epsilon_1 = \sigma/E$ and the permanent strain ϵ_2.

Temperature also has an effect on the stress–strain curves. All elastic moduli decrease with increasing temperature. For example, the temperature coefficient of Young's modulus for iron is $dE/dT = -4\cdot8 \times 10^7 \ \mathrm{N\,m^{-2}\,K^{-1}}$ or, in non-SI units, $-7\cdot0 \times 10^3 \ \mathrm{lbf\,in^{-2}\,K^{-1}}$, over the range $0^\circ - 100^\circ C$. At elevated temperatures, deformation can continue

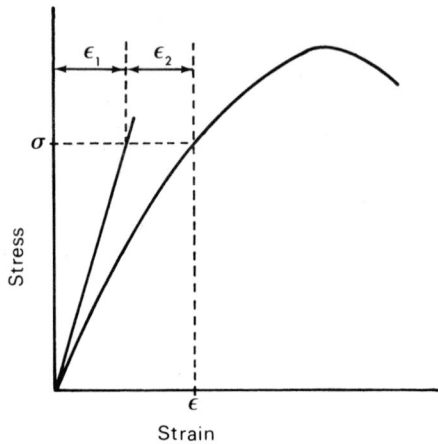

Figure 1.8.   For a ductile material the strain ϵ may be considered to be the sum of the elastic and permanent strains ϵ_1 and ϵ_2

with time even with a constant applied stress. This behaviour is termed *creep*. The deformation takes place at a comparatively slow rate: for example, tests on alloy steels at say $500^\circ C$ may continue over several months or even years. A specimen with a constant applied tensile load of sufficient magnitude to cause creep shows an initial strain consisting of an elastic part and a permanent part. If strain is plotted against time for constant temperature and stress, a plot is obtained of which the first part is curved and is termed the period of primary creep (Figure 1.9). After this the plot becomes straight and the period of secondary creep occurs. This period of constant creep rate may continue for a considerable period, even years, and determines the useful life of the material. The *creep rate* is defined as strain/time over this region. Tertiary creep occurs when the grain boundaries, the regions where the individual and randomly orientated crystallites of the metal join, begin to break down, leading to necking and causing an increase in the slope

10

of the curve. Any increase in temperature or stress will give increased slopes to the curve.

Conversely, if a rod is strained rather than stressed by a constant amount under conditions which would cause creep, a stress will be developed with the deformation. As time passes it will be found that this stress developed will decrease even though the strain is maintained

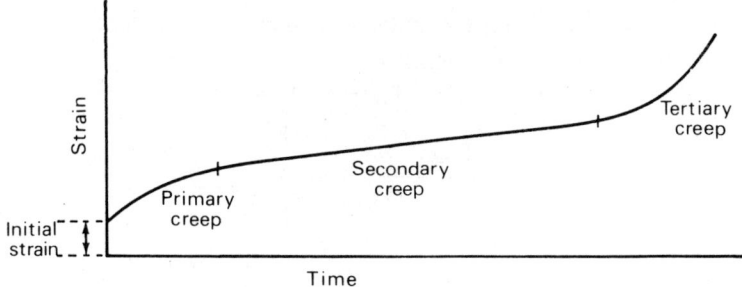

Figure 1.9. Deformation occurring with time for a specimen under a constant tensile stress

constant. This phenomenon is known as *relaxation*. For some materials, such as glass, creep will commence with the smallest of stresses, provided the temperature is high enough, and will continue at a rate that is approximately proportional to the applied stress. The material is then said to be behaving in a *viscous* manner.

When the tensile specimen finally breaks, the nature of the break may take various forms. A sharp break without appreciable plastic deformation or creep is termed a *brittle fracture*, whilst, if there is a gradual thinning down to zero thickness due to plastic or viscous flow, the specimen is said to *rupture*. However, if it breaks before the zero thickness condition is reached, that is, before rupture, the break is said to be a *ductile fracture*. A measure of the ductility of a material is the *percent elongation*, defined as

$$\text{percent elongation} = \frac{l_f - l_0}{l_0} \times 100 \qquad (1.7)$$

where l_0 is the original test length of the specimen and l_f is the final length, determined by reassembly of the fractured rod.

Some indication as to whether a material will be ductile or brittle can also be found from hardness tests, such as the Brinell test. In this a hard steel ball is pressed into the material by a known load, making an indentation of a depth which is a measure of the hardness. It has been found that the Brinell hardness number increases with increasing tensile strength. Other shape indentors are also used — for example, the Vickers test employs a diamond pyramid. For a ductile material there will be a flow of material outwards, tending to pile up at the edge of the indentation. When the load is released, there will be some plastic recovery.

As well as stress—strain curves due to tensile forces, curves due to compressive forces are plotted. These are somewhat similar to the tensile-type curves in that most materials show elastic properties to

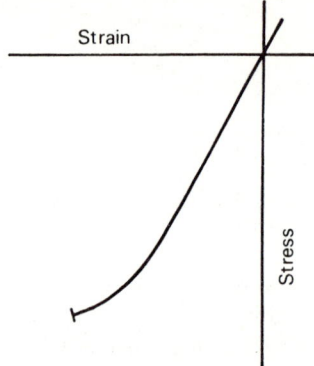

Figure 1.10.   Characteristic shape of the stress—strain curve for a mild steel specimen under compression

some extent giving a linear stress—strain region. Again, as the stress is increased, the elastic limit and a yield point is reached, followed usually by some plastic flow up to the *crushing limit* (as shown in Figure 1.10) when the material completely breaks down.

1.3 Atomic Bonding

To what extent can the results of actual tests be explained? Although it is difficult or even impossible to make accurate predictions of the magnitude of deformations with applied stresses, it is possible to give a reasonable explanation of what is happening when a body is deformed and to make qualitative predictions. To do this we must make recourse to knowledge of the bonding forces between atoms. It may be recalled that the strongest bonds are the ionic, covalent, and metallic bonds.

12

An *ionic bond* occurs between atoms with easily detachable valence electrons, such as the alkali metals, and atoms with a tendency to acquire electrons to form a complete outer electron shell. Such elements are the halogens. For example, a neutral sodium atom (atomic number 11) will readily become positively ionised by losing its single 3s valence electron and so attain the stable electron configuration of neon (atomic number 10). Similarly, chlorine (atomic number 17) will readily become negatively ionised by acquiring an electron to complete its outer 3p group and so form the stable electron configuration of argon (atomic number 18). In other words, an electron from a neutral sodium atom leaves it and joins onto a neutral chlorine atom, so the chlorine atom acquires a negative charge and becomes a negative ion, and the sodium atom having lost a negative charge becomes a positive ion. Positive sodium ions and negative chlorine ions are then able to attract each other electrostatically. There is, however, a limit to how close the ions can approach each other. If they become too close, their outer full electron shells would start to overlap. This is not allowed by the Pauli exclusion principle, and the ions must consequently repel each other by a force depending on the amount of overlap. The ions will take up an equilibrium position where this repulsion force and that due to the electrostatic attraction are balanced, to form a giant molecule of solium chloride, as shown in Figure 1.11.

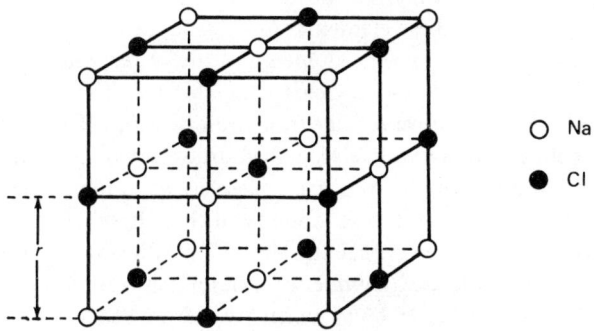

○ Na

● Cl

Figure 1.11. Sodium chloride structure

The *covalent bond* cannot be explained quite so simply, but it arises from a tendency of certain elements to share their valence electrons. It can be shown by wave mechanics that this sharing leads to a reduction in energy of the combined system and consequently a binding force. For example, in a methane molecule CH_4, the four

13

valence electrons of the carbon atom each pair up with the electron of each of the four hydrogen atoms. The resulting molecule has a tetrahedral arrangement of four lobes each containing two electrons, with the carbon atom at the centre of the system and a hydrogen atom at the end of each lobe (Figure 1.12). The tetrahedral arrangement

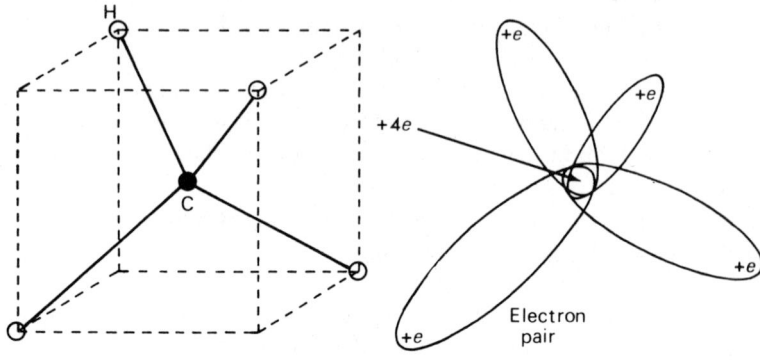

Figure 1.12. Methane molecule

results from the mutual electrostatic repulsion of the negatively charged lobes.

The *metallic bond* is characteristic only of the metals, including the alkali metals, and results from the easily detachable nature of the valence electrons of these elements. Again from wave mechanics it may be shown there is a reduction of energy of the system and hence a binding force, if the atoms can pack closer together by allowing their valence electrons to move readily from atom to atom. These valence electrons thus form a negatively charged 'free-electron gas', pervading the crystal and holding the positively charged metal ions together in a close-packed arrangement. The mobility of these 'free' electrons accounts for the high electrical and thermal conductivity of the metals.

In addition, there are a number of other much weaker binding forces, such as the *van der Waals' bond* which arises from oscillations in the electron clouds representing an oscillating dipole. These oscillations synchronise with those of neighbouring electron clouds to produce other dipoles which attract each other. Of course, several different bonding mechanisms may be in operation simultaneously in a particular instance.

Returning now to the problem of the explanation of elasticity, it will readily be seen that a stretching of these bonds could lead on a macroscopic scale to the elastic portion of the stress–strain curves of the previous section, whilst too much stretching could cause a breaking of the bonds and give the non-linear plastic-flow region of the curves. In other words, if the bonds can be imagined as springs, a tensile force stretches the springs and causes the atoms to move farther apart and a compressive one forces the atoms closer together. This spring concept can be examined more closely by a consideration of the bonding of an ionic compound, for example a crystal of sodium chloride. Ionic bonding is chosen as an example simply because the other types of bond lead to less easily resolvable complexities, as do the more complicated crystal structures.

Let us now consider a crystal of sodium chloride as a cubic array of negative and positive ions, as in Figure 1.11, with each ion having a spherical distribution of charge and occupying its correct lattice position. This is a slight simplification of the true case since, for example, there would normally be at least some lattice defects due to impurities or to vacant lattice sites. However, in this idealised model the force between ions depends solely on their distance apart. The Coulomb force between two ions, one of charge $Z_1 e$ and the other of charge $Z_2 e$ and distance r apart, is

$$F_C = - \frac{(Z_1 e)\,(Z_2 e)}{4\pi\epsilon_0 r^2} \tag{1.8}$$

and thus the force of attraction between a monovalent positive sodium ion and a monovalent negative chlorine ion distance r apart is

$$F_C = \frac{e^2}{4\pi\epsilon_0 r^2} \tag{1.9}$$

where e is the charge on each ion and ϵ_0 is the permittivity of free space, approximately $8.85 \times 10^{-12}\ \mathrm{F\,m^{-1}}$. Generally the charge e represents the product of the charge and valency of the ions. The coefficient $1/4\pi\epsilon_0$ arises from the choice of units.

The attractive energy between these two ions due to the Coulomb force may then be calculated since the energy is the product of force and distance, that is,

$$E_C = \int F_C \, \mathrm{d}r \tag{1.10}$$

It is convention to assume that this energy will be zero when the ions are an infinite distance apart, that is, $E_C = 0 = E_\infty$ when $r \to \infty$. The attractive energy for these two ions then is

15

$$E_C = \int_{\infty}^{r} \frac{e^2}{4\pi\epsilon_0 r^2} \, dr$$

$$= -\frac{e^2}{4\pi\epsilon_0 r} \tag{1.11}$$

However, we must take into account the effects of all the ions within the crystal. From Figure 1.11, it will be seen that, if the shortest interionic distance is r, every positive sodium ion is attracted by six negative chlorine ions at a distance r, eight at a distance $r\sqrt{3}$, etc., at the same time that it is repelled by twelve positive sodium ions at a distance $r\sqrt{2}$, six at a distance $2r = r\sqrt{4}$, etc. Thus the attractive energy due to the forces between the monovalent positive sodium ion and all the surrounding monovalent ions will be

$$E_C = -\frac{e^2}{4\pi\epsilon_0 r} \left(\frac{6}{\sqrt{1}} - \frac{12}{\sqrt{2}} + \frac{8}{\sqrt{3}} - \frac{6}{\sqrt{4}} + \frac{24}{\sqrt{5}} - \frac{24}{\sqrt{6}} + \cdots \right)$$

$$= -\frac{Me^2}{4\pi\epsilon_0 r}$$

The coefficient M is termed the *Madelung constant* and is the pure number sum of the series, which depends only on the crystal structure. It is not an easy sum to evaluate as the series is only very slowly convergent. For sodium chloride, M has the value 1·7476. It has been determined for a number of other ionic crystals: for example, the Madelung constant for fluorite (CaF_2) is 5·0388 and for corundum (Al_2O_3) 25·0312. The attractive energy can thus be written as

$$E_C = -\frac{Ae^2}{r}$$

where A is a constant equal to $M/4\pi\epsilon_0$. If there are N such positive ions in the crystal, the total energy due to attraction will be

$$E_C = -\frac{ANe^2}{r} \tag{1.12}$$

As well as an attraction due to electrostatic forces, there must also be a repulsive force otherwise the ions would collapse onto each other. If the ions approach too close to each other, the outer full electron shells would begin to overlap, contravening the Pauli exclusion principle. This in turn constitutes a repulsive force whose magnitude depends on the amount of overlap of the wave functions of the

16

valence electrons. The repulsive force exerted on the positive ion by all the surrounding ions may be expressed as

$$F_R = -\frac{nB}{r^{n+1}} \qquad (1.13)$$

where n and B are constants; B is a constant depending on the lattice configuration and n is approximately equal to 9 for ionic solids. More accurately it may be shown that the repulsive force falls off exponentially. However, it will be seen that, owing to the high value of the exponent, the repulsive force falls off far more rapidly than the force of attraction.

We may now calculate the repulsive energy due to this repulsive force on the positive ion since, as before,

$$E_R = -\int_{\infty}^{r} \frac{nB}{r^{n+1}} \, dr = \frac{B}{r^n} \qquad (1.14)$$

Again, if there are N such positive ions in the crystal, the total repulsive energy will be

$$E_R = \frac{BN}{r^n} \qquad (1.15)$$

Thus the binding energy of the crystal, that is, the binding energy for a crystal containing N pairs of positive and negative ions with each ion being attracted and repelled by all the surrounding ions, is

$$E = N\left(\frac{B}{r^n} - \frac{Ae^2}{r}\right) \qquad (1.16)$$

from equations 1.12 and 1.15. In Figure 1.13(a) this energy, as well as the separate energies of attraction and repulsion, are plotted as functions of the distance r. With a separation of r_0, the total energy will be a minimum and r_0 becomes the equilibrium interatomic distance. This is the distance where the force of attraction F_C becomes equal in magnitude to the force of repulsion F_R, as is shown in Figure 1.13(b).

Returning again to the problem of elasticity, it is easy to see that, if a tensile stress is applied to the crystal, the interatomic distance r will be increased. This will raise the energy of the system and bring into play the attractive force which will try to return the interatomic distance r to its equilibrium value r_0. Similarly, a compressive stress will bring into play the repulsive force with the same end in mind.

The sum of the attractive and repulsive forces for a particular interatomic distance r will be proportional to the stress that must be applied

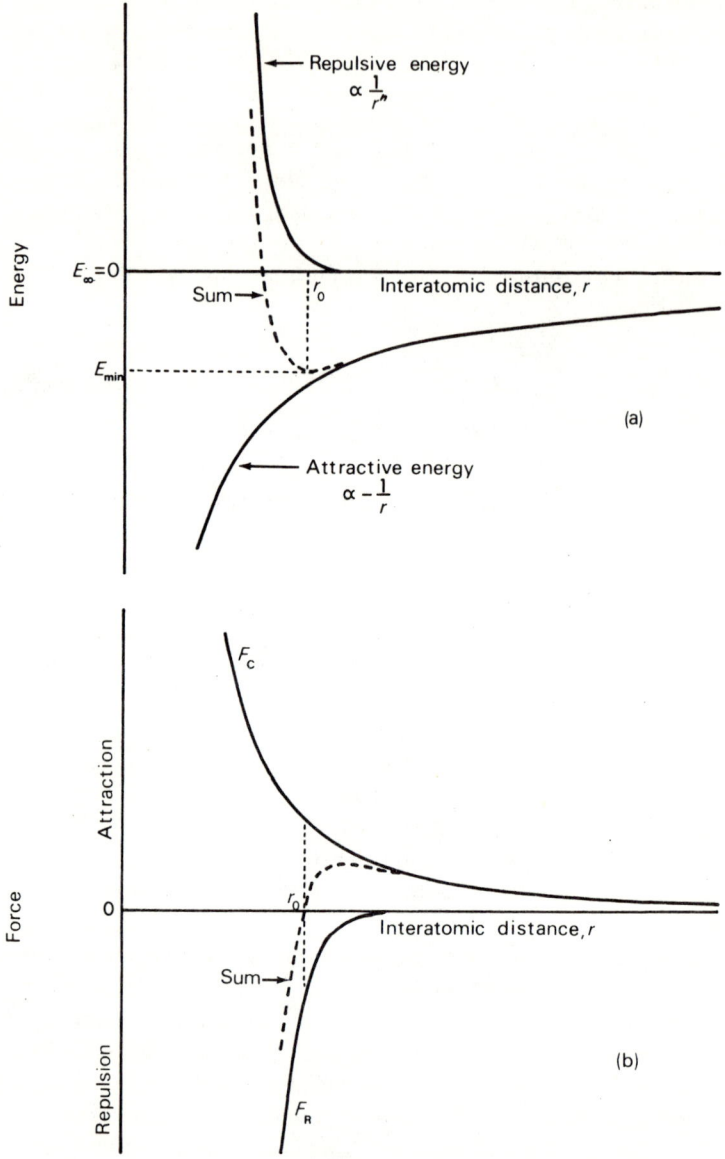

Figure 1.13.   (a) Energies and (b) forces involved in the bonding of
atoms of an ionic crystal, plotted as functions of the
interatomic distance r

to change that distance from its equilibrium distance r_0 to the value r, that is, by an amount Δr. Thus, if the net forces are plotted against $\Delta r/r_0$, a curve analogous to the stress–strain curves of the previous section will be obtained, at least for small strains, showing that the theory is in approximate agreement with Hooke's law. For larger strains, however, the asymmetry about zero of the curve of the sum of the attractive and repulsive forces as shown in Figure 1.13(b) indicates a departure from Hooke's law. The theory we have just discussed is then not sufficient to explain the observed linearity between stress and strain, and it becomes necessary to consider the effects of the movements of dislocations.

Similar treatments may be made of the other types of crystal bonding and for more complicated crystal structures, although as was said earlier such treatments may be more difficult. Generally, however, the results give approximate agreement with Hooke's law and explain satisfactorily the phenomenon of elasticity for small deformations. Again as was said earlier, larger deformations involve plastic flow, a discussion of which is deferred until Chapter 3.

1.4 Stress and Strain

In Section 1.1 we discussed the meanings of stress and strain in general terms. We need now to be more specific: for example, if a stress is applied to a body in a particular direction to produce a strain in that direction, there will also be other stresses and strains induced in various other directions. We need to be able to describe all the directions and magnitudes of these stresses and strains in a precise and simple manner. To do this we make use of tensors, which will be defined as we go along. These have a great advantage in theoretical studies because the parameters can be so precisely defined and facilitate the setting up of computer programmes for the solution of problems. There is also a well-established algebra of tensors so that, for example, axes may be readily changed, although this will not concern us here. Many other physical properties, such as thermal and electrical conductivities, and piezoelectricity, can also be described by means of tensors, and it is because of their general usefulness in physics that they will be used here in the discussion of stress and strain.

In the treatment of stress and strain in practical cases, it is necessary to deal with an element of such a size that the effects of individual atoms have been averaged out; at the same time, however, the element must not be too large otherwise the problem may become too complex owing to large stress gradients with their complementary strain gradients occurring over the element. Therefore we must bear in mind the proviso that an element of suitable dimensions is now being considered.

Let us deal first with stress and consider the stresses acting on an elemental cube inside a body that is being subjected to tensile forces. In Figure 1.14 A_1, A_2, A_3, A_{-1}, A_{-2}, and A_{-3} represent the equal

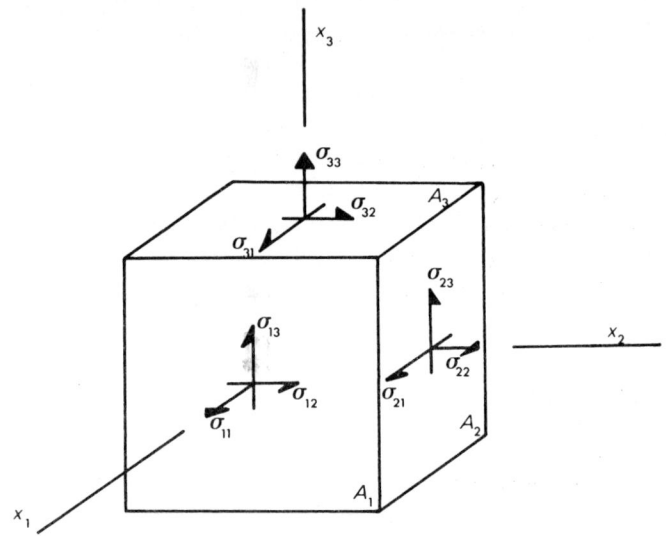

Figure 1.14. Components of stress acting on a cube

areas of the six faces of a cube element on which forces are acting in arbitrary directions. The force on each face may be resolved into three components acting parallel to the x_1, x_2, and x_3 axes, so that the stress, for example, on the face of area A_1 (that is, the face whose outward normal is in the direction x_1) may be written as

$$\sigma_{11} = F_1/A_1 \tag{1.17a}$$
$$\sigma_{12} = F_2/A_1 \tag{1.17b}$$
$$\sigma_{13} = F_3/A_1 \tag{1.17c}$$

where F_1, F_2, and F_3 are the three components in the x_1, x_2, and x_3 directions of the force acting on this face. The first subscript of the stress σ denotes the direction of the outward normal to the surface,

20

and the second subscript denotes the particular component of the force on that surface. If both subscripts have the same signs, as they have in this case, then the stress will be a tensile one. If the subscripts have opposite signs, then compressive forces will be involved.

The components σ_{11}, σ_{22}, and σ_{33} are termed the *normal components* of stress, and σ_{12}, etc., are the *shear components*. Altogether there will be 18 stress components acting on the cube, but fortunately most of these are equal. For statical equilibrium of the cube element and from considerations of symmetry, the forces on the other three faces of the cube must be equal and opposite to those shown in Figure 1.14, that is, $\sigma_{11} = \sigma_{-1-1}$, $\sigma_{23} = \sigma_{-2-3}$, etc., reducing the number of stress components to nine. These may be written as an array:

$$\begin{bmatrix} \sigma_{11} & \sigma_{12} & \sigma_{13} \\ \sigma_{21} & \sigma_{22} & \sigma_{23} \\ \sigma_{31} & \sigma_{32} & \sigma_{33} \end{bmatrix} \qquad (1.18)$$

Also for statical equilibrium, there must be no moments about each of the axes x_1, x_2, and x_3. Thus for zero moments about x_1, as in Figure 1.15,

$$\sigma_{23} = \sigma_{32}$$

and generally

$$\sigma_{ij} = \sigma_{ji} \qquad (1.19)$$

It will thus be seen that the array of nine stress components is symmetrical about the principal diagonal, and the nine components consist of six independent ones. The nine parameters form an array known as a *tensor*.

A scalar quantity is termed a zero-rank tensor and requires one number to specify it; a vector is termed a first-rank tensor and requires three numbers related to the three reference axes to specify it; a second-rank tensor, such as stress, requires nine numbers to specify it, each number being associated with a pair of axes. In addition there are higher-rank tensors — for example, fourth-rank tensors are used in relating stress to strain. A tensor may be thought of as representing a physical quantity that is independent of the choice of axes. If the axes are changed, the representation of the physical quantity changes but not the physical quantity itself. A second-rank tensor really relates two vectors; in our case the force vector is related to the vector denoting area.

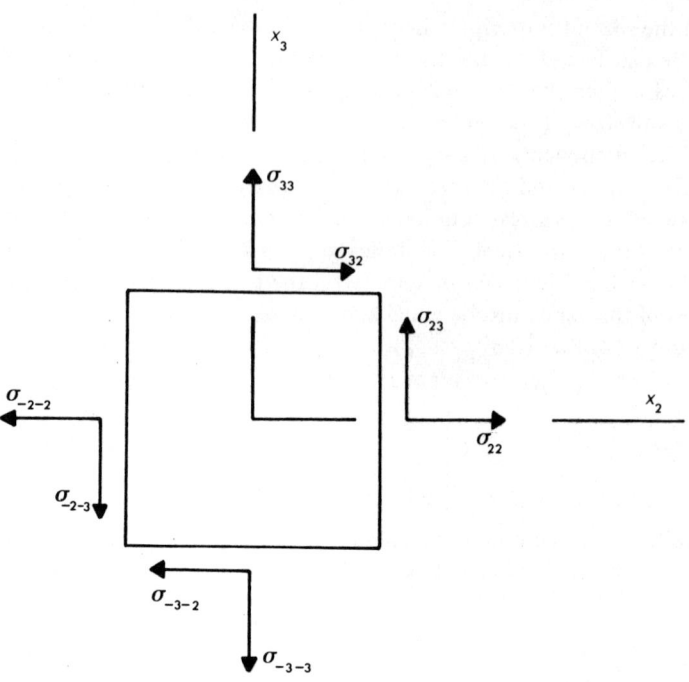

Figure 1.15. Normal and shear components of stress having moments about the x_1 axis of a cube

The stress tensor so far discussed is the one that arises when forces are applied in general directions to a volume element. If the force direction is restricted (for example, a vertical wire or rod stressed longitudinally by a hanging weight), the stress tensor simplifies to

$$\begin{bmatrix} \sigma & 0 & 0 \\ 0 & 0 & 0 \\ 0 & 0 & 0 \end{bmatrix} \qquad (1.20)$$

For a body subjected to a hydrostatic pressure p, producing compression of the body, the stress tensor becomes

$$\begin{bmatrix} -p & 0 & 0 \\ 0 & -p & 0 \\ 0 & 0 & -p \end{bmatrix} \qquad (1.21)$$

since there are no shear stress components.

22

Strain may also be written as a tensor although it is more difficult
to formulate. Here we are concerned with the relative displacements
between different points in the body, not their actual displacements
which obviously depend on their original separations. The difficulty
arises in separating out the effects of translational movements or
rotations of the whole body from the true strain. To see how this may
be done, let us consider a two-dimensional elemental rectangular body
ABCD on which is marked an element PQ of length δx_1, as shown in
Figure 1.16. If this body is now deformed by a small amount to

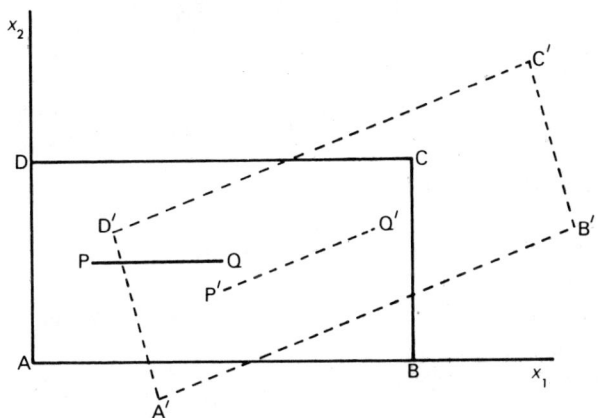

Figure 1.16. Displacement of element PQ to PQ' due to deformation
of two-dimensional elemental body ABCD to A'B'C'D'

A'B'C'D', the line PQ will be displaced to P'Q'. Point Q being further
from the origin than point P will consequently be displaced by a
relatively greater amount than P. Let now the displaced body be translated linearly so that point P' is coincident with point P, as in Figure
1.17. We are now able to consider the strain produced at the point P
without the added confusion due to whole-body movements. The
length of PQ was δx_1. It is deformed to length PQ' by a displacement
of Q by an amount δu_1 in the x_1 direction and an amount δu_2 in the
x_2 direction, where both δu_1 and δu_2 are very much smaller than δx_1.
The actual displacement of Q to Q' will be a function of its original
distance from P, that is, a function of the distance δx_1. Then the
displacement δu_1 in the x_1 direction will be

$$\delta u_1 = \frac{\partial u_1}{\partial x_1} \delta x_1 \qquad (1.22)$$

23

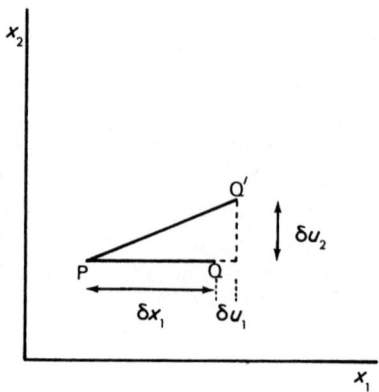

*Figure 1.17. Position of Q' after linear translation of body so that
P' is coincident with P*

and similarly in the x_2 direction

$$\delta u_2 = \frac{\partial u_2}{\partial x_1} \delta x_1 \qquad (1.23)$$

Thus the strain at P, measured in the x_1 direction, will be

$$\frac{\text{increase in length}}{\text{original length}} = \frac{(\partial u_1/\partial x_1) \delta x_1}{\delta x_1} = \frac{\partial u_1}{\partial x_1} \qquad (1.24)$$

and similarly the strain in the x_2 direction will be

$$\frac{(\partial u_2/\partial x_1) \delta x_1}{\delta x_1} = \frac{\partial u_2}{\partial x_1} \qquad (1.25)$$

If we now consider the more general case of a three-dimensional body with the line PQ not simply parallel to x_1 but inclined at some general angle in space, then displacements δu_1, δu_2, and δu_3 respectively measured in the x_1, x_2, and x_3 directions will be involved. The strains may then be written as an array of nine partial derivatives, as follows:

$$
\begin{bmatrix}
\dfrac{\partial u_1}{\partial x_1} & \dfrac{\partial u_1}{\partial x_2} & \dfrac{\partial u_1}{\partial x_3} \\[2ex]
\dfrac{\partial u_2}{\partial x_1} & \dfrac{\partial u_2}{\partial x_2} & \dfrac{\partial u_2}{\partial x_3} \\[2ex]
\dfrac{\partial u_3}{\partial x_1} & \dfrac{\partial u_3}{\partial x_2} & \dfrac{\partial u_3}{\partial x_3}
\end{bmatrix}
=
\begin{bmatrix}
e_{11} & e_{12} & e_{13} \\[2ex]
e_{21} & e_{22} & e_{23} \\[2ex]
e_{31} & e_{32} & e_{33}
\end{bmatrix}
\qquad (1.26)
$$

24

This array is still not sufficient, however, to describe completely the strain as it does not preclude the possibility of rotation. The displacement of PQ to PQ′ could have been achieved by a rotation plus a tensile extension. To examine the implications of this, let us again consider the displacement of the elemental line PQ on the surface of a two-dimensional body to position P′Q′, as shown in Figure 1.16. As well as this line, let us also consider the displacement of another elemental line PR parallel to the x_2 axis. On deformation of the rectangle ABCD to A′B′C′D′, line PR will be displaced to P′R′. The whole body may then be linearly translated so that point P′ is coincident with P as was done previously. The lengths and orientations of the two lines, before and after deformation, are shown in Figure 1.18. For a small rotation of

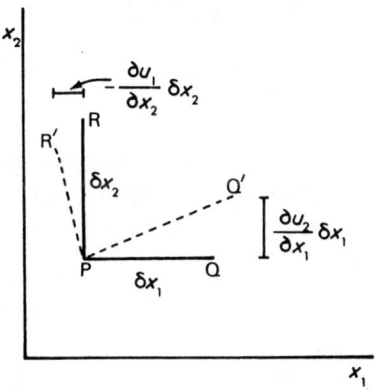

Figure 1.18. Position of Q′ and R′ after linear translation of body so that P′ is coincident with P

PQ, initially parallel to the x_1 direction, the angle of rotation in radians is given by

$$\omega = \frac{\partial u_2}{\partial x_1} = e_{21} \tag{1.27}$$

where e_{21} denotes a rotation about the x_3 axis towards the x_2 axis of a line initially parallel to the x_1 axis.

Similarly, the rotation of PR, initially parallel to the x_2 direction, is

$$\omega = -\frac{\partial u_1}{\partial x_2} = -e_{12} \tag{1.28}$$

It is negative because the displacement of R to R′ involves a movement in the negative direction of the x_1 axis.

25

The average rotation in the $x_1 x_2$ plane may be defined as

$$\omega = \frac{1}{2}\left(\frac{\partial u_2}{\partial x_1} - \frac{\partial u_1}{\partial x_2}\right) = \frac{1}{2}\left(e_{21} - e_{12}\right) \quad (1.29)$$

Again this may be generalised for the three-dimensional problem and involves an array of nine as before:

$$\begin{bmatrix} 0 & \frac{1}{2}\left(\frac{\partial u_1}{\partial x_2} - \frac{\partial u_2}{\partial x_1}\right) & \frac{1}{2}\left(\frac{\partial u_1}{\partial x_3} - \frac{\partial u_3}{\partial x_1}\right) \\ \frac{1}{2}\left(\frac{\partial u_2}{\partial x_1} - \frac{\partial u_1}{\partial x_2}\right) & 0 & \frac{1}{2}\left(\frac{\partial u_2}{\partial x_3} - \frac{\partial u_3}{\partial x_2}\right) \\ \frac{1}{2}\left(\frac{\partial u_3}{\partial x_1} - \frac{\partial u_1}{\partial x_3}\right) & \frac{1}{2}\left(\frac{\partial u_3}{\partial x_2} - \frac{\partial u_2}{\partial x_3}\right) & 0 \end{bmatrix} \quad (1.30)$$

The diagonal terms involve $\partial u_1/\partial x_1$, etc., and imply a rotation towards an axis of a line that is already parallel to that axis — a rotation that is obviously zero.

It is now only necessary to subtract from the strains of array 1.26 the effects of any rotation, as given by array 1.30, to give an array of true strains which completely describes the local changes in shape occurring in a body:

$$\begin{bmatrix} \epsilon_{11} & \epsilon_{12} & \epsilon_{13} \\ \epsilon_{21} & \epsilon_{22} & \epsilon_{23} \\ \epsilon_{31} & \epsilon_{32} & \epsilon_{33} \end{bmatrix}$$

$$= \begin{bmatrix} \frac{\partial u_1}{\partial x_1} & \frac{1}{2}\left(\frac{\partial u_1}{\partial x_2} + \frac{\partial u_2}{\partial x_1}\right) & \frac{1}{2}\left(\frac{\partial u_1}{\partial x_3} + \frac{\partial u_3}{\partial x_1}\right) \\ \frac{1}{2}\left(\frac{\partial u_2}{\partial x_1} + \frac{\partial u_1}{\partial x_2}\right) & \frac{\partial u_2}{\partial x_2} & \frac{1}{2}\left(\frac{\partial u_2}{\partial x_3} + \frac{\partial u_3}{\partial x_2}\right) \\ \frac{1}{2}\left(\frac{\partial u_3}{\partial x_1} + \frac{\partial u_1}{\partial x_3}\right) & \frac{1}{2}\left(\frac{\partial u_3}{\partial x_2} + \frac{\partial u_2}{\partial x_3}\right) & \frac{\partial u_3}{\partial x_3} \end{bmatrix}$$

$$= \begin{bmatrix} e_{11} & \frac{1}{2}\left(e_{12} + e_{21}\right) & \frac{1}{2}\left(e_{13} + e_{31}\right) \\ \frac{1}{2}\left(e_{21} + e_{12}\right) & e_{22} & \frac{1}{2}\left(e_{23} + e_{32}\right) \\ \frac{1}{2}\left(e_{31} + e_{13}\right) & \frac{1}{2}\left(e_{32} + e_{23}\right) & e_{33} \end{bmatrix} \quad (1.31)$$

This may be written briefly as

$$\epsilon_{ij} = \frac{1}{2}(e_{ij} + e_{ji}) \qquad (1.32)$$

where i and $j = 1, 2, 3$. The diagonal terms e_{11}, e_{22}, and e_{33} are the strains measured in the directions of the x_1, x_2, and x_3 axes respectively and are termed the *tensile strains*. The other terms involving e_{ij}, where $i \neq j$, imply rotations as already defined and measure the *shear strains*.

It will also be seen that the array is symmetrical about the principal diagonal, so $\epsilon_{ij} = \epsilon_{ji}$ and only six of the terms out of the nine are independent. Strain is thus similar to stress in that it forms a symmetrical second-rank tensor of six independent terms, three representing tensile effects and three shear effects. It is a second-rank tensor because it relates two vectors, one representing the difference in displacement of two points and the other representing their original separation.

Supposing now that a square body is deformed by a tensile force, as shown in Figure 1.19; then two sides will be rotated by e_{21} and e_{12} as

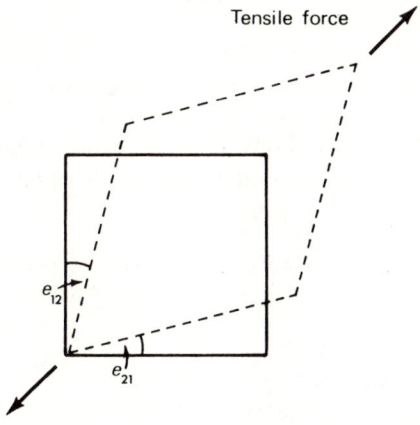

Figure 1.19. Deformation of a square body by a tensile force

shown and the rotations are as defined by equations 1.27 and 1.28. The total shear angle $(e_{12} + e_{21})$ is equal to the angle of shear γ, as defined by equation 1.5 and Figure 1.1, where $\tan \gamma = \gamma$ for small shear strains. To avoid ambiguity, γ may be termed the *engineering shear strain* to distinguish it from ϵ_{ij}, the *tensor shear strains*. Thus

$$e_{12} + e_{21} = \gamma_{12} \qquad (1.33)$$

27

The subscript to γ implies that γ_{12} is the decrease in angle between two lines originally parallel to the x_1 and x_2 directions. But, from equation 1.31,

$$\epsilon_{12} = \frac{1}{2}(e_{12} + e_{21}) = \frac{1}{2}\gamma_{12} \qquad (1.34)$$

and generally

$$\epsilon_{ij} = \frac{1}{2}\gamma_{ij} \qquad (1.35)$$

That is, the tensor shear strain is half the engineering shear strain occurring in the definition of the shear modulus of rigidity. In other words, it is half the change in angle between two originally mutually perpendicular lines in the body. Thus, if two lines drawn in the body are initially parallel to the x_1 and x_2 axes, then after deformation of the body the angle between them will be $\frac{1}{2}\pi - 2\epsilon_{12}$.

We have already stated Hooke's law in a simple form (equations 1.1 and 1.2):

$$\epsilon = s\,\sigma$$

where s was the compliance and equal to $1/c$, and c was the stiffness constant. We can now extend this by making use of the new notation which defines stress and strain more precisely. If a general stress is applied to a homogeneous body, then a resulting component of strain ϵ_{ij} will be related to all nine components of the applied stress. For example,

$$\begin{aligned}
\epsilon_{11} = \quad & s_{1111}\,\sigma_{11} + s_{1112}\,\sigma_{12} + s_{1113}\,\sigma_{13} \\
+ & s_{1121}\,\sigma_{21} + s_{1122}\,\sigma_{22} + s_{1123}\,\sigma_{23} \\
+ & s_{1131}\,\sigma_{31} + s_{1132}\,\sigma_{32} + s_{1133}\,\sigma_{33}
\end{aligned} \qquad (1.36)$$

and similarly for the other eight components of the strain — a total of nine equations and 81 s-constants. The generalised form of Hooke's law may be written as

$$\epsilon_{ij} = s_{ijkl}\sigma_{kl} \qquad (1.37)$$

where i, j, k, and $l = 1, 2, 3$. There are consequently 81 corresponding stiffness constants c_{ijkl} which are linear functions of s_{ijkl}. Equation 1.36 also implies that a single component of stress (for example, a body being stretched by a uniaxially applied tensile force), may possibly bring in all the components of strain, including shear. The 81 compliance constants and the 81 stiffness constants each form a fourth-rank tensor.

Luckily, owing to symmetry considerations that, for example, made $\epsilon_{ij} = \epsilon_{ji}$, the 81 constants can be reduced to 36 independent terms since $s_{ijkl} = s_{jikl}$ and $c_{ijkl} = c_{jikl}$. In specific problems, the number of independent terms may be reduced even further but will depend on the crystallographic structure and the relationship of the chosen axes with the crystallographic axes. It may be seen that a full definition of Hooke's law is possible using tensor notation, enabling the deformation of a body to be described completely.

In practical problems it may be possible to reduce the number of independent stress—strain parameters involved so that Young's, bulk, and shear moduli and Poisson's ratio are the main properties to be determined (equations 1.3—1.6). It is usual to measure these directly but even so it may not be possible or convenient to measure them all, in which case an unknown one must be calculated from a knowledge of the others. This is possible since, as we have seen above, a component of stress may produce components of all the strains and thus the application of a particular stress and the measurement of a particular strain, as required in the measurement of a modulus, must produce other stresses and strains which are involved with a different modulus. Thus a body may be deformed by a particular amount by first changing the size but not the shape, which involves the bulk modulus, and then by changing the shape but not the size, which involves the shear modulus. In the next section, therefore, we shall determine the relationships between the moduli and Poisson's ratio.

1.5 Relationships between the Moduli

When measurements of the moduli are made, they are made in a particular way that avoids unwanted components. Thus, if we consider a cube with edges parallel to the x_1, x_2, and x_3 axes and apply stresses only in directions parallel to these axes, only strains in these directions need be considered as any shear strains will be negligible in comparison. Neglecting then the shear terms, equation 1.36 simplifies to

$$\epsilon_{11} = s_{1111}\, \sigma_{11} + s_{1122}\, \sigma_{22} + s_{1133}\, \sigma_{33} \qquad (1.38)$$

Since the subscripts now occur only in pairs, we may simplify the notation slightly by writing equation 1.38 and the remaining equations involving only normal stresses as

$$\epsilon_1 = s_{11}\,\sigma_1 + s_{12}\,\sigma_2 + s_{13}\,\sigma_3 \qquad (1.39a)$$

$$\epsilon_2 = s_{21}\,\sigma_1 + s_{22}\,\sigma_2 + s_{23}\,\sigma_3 \qquad (1.39b)$$

$$\epsilon_3 = s_{31}\,\sigma_1 + s_{32}\,\sigma_2 + s_{33}\,\sigma_3 \qquad (1.39c)$$

Now from the definition of Young's modulus (equation 1.3),

$$s_{11} = \frac{\epsilon_1}{\sigma_1} = \frac{1}{E} \qquad (1.40)$$

measured in the same direction as the applied stress σ_1 and resulting strain ϵ_1 and where E is Young's modulus. Also, for an applied stress σ_1 only, that is, one applied in only the x_1 direction,

$$s_{21} = \frac{\epsilon_2}{\sigma_1} = \frac{\epsilon_2/\epsilon_1}{\sigma_1/\epsilon_1} = -\frac{\nu}{E} \qquad (1.41)$$

from equations 1.39 and the definition of Poisson's ratio (equation 1.6).

For an isotropic material, a given stress applied in the x_1, x_2, or x_3 direction will always produce the same strain measured in that direction. In other words, the constant of proportionality will be the same for all strains measured in the directions of their applied stresses, so

$$s_{11} = s_{22} = s_{33} \qquad (1.42)$$

Similarly, there will be the same constant of proportionality between all stresses and their transverse produced strains, so

$$s_{21} = s_{31} = s_{12} = s_{32} = s_{13} = s_{23} \qquad (1.43)$$

Therefore we can rewrite equation 1.39 in terms of Young's modulus and Poisson's ratio:

$$\epsilon_1 = \frac{1}{E}[\sigma_1 - \nu\,(\sigma_2 + \sigma_3)] \qquad (1.44a)$$

$$\epsilon_2 = \frac{1}{E}[\sigma_2 - \nu\,(\sigma_1 + \sigma_3)] \qquad (1.44b)$$

$$\epsilon_3 = \frac{1}{E}[\sigma_3 - \nu\,(\sigma_1 + \sigma_2)] \qquad (1.44c)$$

These three simultaneous equations may now be solved for stresses σ_1, σ_2, and σ_3. This may be done by, for example, adding the second and third equations and substituting $\sigma_2 + \sigma_3$ from the first equation to obtain

$$\sigma_1 = \frac{E\nu(\epsilon_2 + \epsilon_3 - \epsilon_1) + E\epsilon_1}{(1 + \nu)(1 - 2\nu)}$$

$$= \frac{E\nu(\epsilon_1 + \epsilon_2 + \epsilon_3) + E(\epsilon_1 - 2\epsilon_1\nu)}{(1 + \nu)(1 - 2\nu)}$$

$$= \frac{E\nu(\epsilon_1 + \epsilon_2 + \epsilon_3)}{(1 + \nu)(1 - 2\nu)} + \frac{E\epsilon_1}{(1 + \nu)} \tag{1.45}$$

and similarly for σ_2 and σ_3. It is convenient to make the following substitutions in these equations:

$$\Delta = \epsilon_1 + \epsilon_2 + \epsilon_3 \tag{1.46}$$

$$n = \frac{E\nu}{(1 + \nu)(1 - 2\nu)} \tag{1.47}$$

$$2G' = \frac{E}{1 + \nu} \tag{1.48}$$

The equations for normal stresses σ_1, σ_2, and σ_3 may now be written as

$$\sigma_1 = n\Delta + 2G'\epsilon_1 \tag{1.49a}$$

$$\sigma_2 = n\Delta + 2G'\epsilon_2 \tag{1.49b}$$

$$\sigma_3 = n\Delta + 2G'\epsilon_3 \tag{1.49c}$$

Δ is termed the *dilation* and measures the fractional change in volume. It will be shown later that $G' = G$, the shear modulus.

Considering now the bulk modulus K, this has been defined by equation 1.4 as

$$K = \frac{\text{bulk stress}}{\text{bulk strain}}$$

$$= \frac{\text{pressure}}{\text{change in volume per unit volume (dilation)}} \tag{1.50}$$

Also the pressure p in this case will be equal to the applied stress on each face of the cube, so $\sigma_1 = \sigma_2 = \sigma_3 = p$ and hence adding the three equations 1.49 together gives

$$\Delta = \frac{3p}{3n + 2G'} \tag{1.51}$$

Therefore, on substituting from equations 1.47, 1.48, and 1.51,

$$K = \frac{p}{\varDelta}$$

$$= \frac{3n + 2G'}{3}$$

$$= \frac{E}{3(1 - 2\nu)} \tag{1.52}$$

which relates the bulk modulus K, Young's modulus E, and Poisson's ratio ν.

To show that G' is equal to the shear modulus G requires a short preliminary derivation. We have to show that a shearing stress applied to the cube corresponds to a tensile stress of the same magnitude along one diagonal and an equal compressive stress along the other diagonal.

Consider a shearing force F acting along face AB of a cube of face area A, as shown in Figure 1.20. This will produce a shearing stress

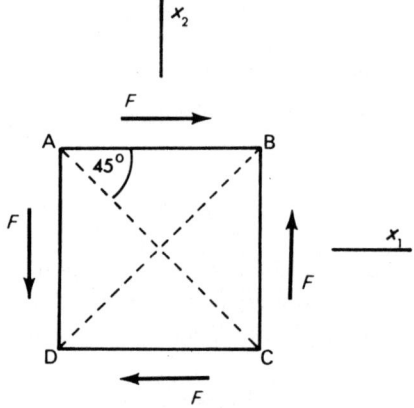

Figure 1.20.   Action of a shearing force F along the face of a cube

equal to F/A. There are no stresses applied in the x_3 direction. For equilibrium, the other forces shown will also be brought into play. The two forces acting along AB and AD may be resolved into one force, $2F \sin 45° = F\sqrt{2}$, acting along the diagonal AC. There will be an equal force acting along CA due to the forces along CB and CD. Thus there will be a compressive force $F\sqrt{2}$ acting on the diagonal plane DB. The area of this plane is $A/\sin 45° = A\sqrt{2}$ and the compressive stress acting along the diagonal AC is therefore $(F\sqrt{2})/(A\sqrt{2}) = F/A$.

32

There will be similarly an equal tensile stress acting along the diagonal
BD. That is, the applied shearing stress is equivalent to an equal tensile
stress along one diagonal and an equal compressive stress along the
other.

Returning now to the derivation of an expression for the shear
modulus, the stresses and strains involved are shown in Figure 1.21(a).

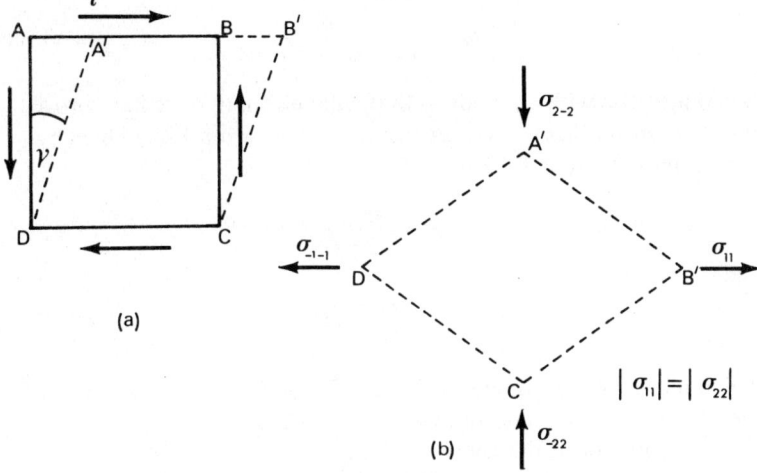

(a)

(b)

$|\sigma_{11}| = |\sigma_{22}|$

Figure 1.21. Deformation of a cube (a) by a shear stress, and
(b) by equal tensile and compressive stresses

Here a shear stress τ is producing a small shear strain γ (equal to
$\tan \gamma$ where γ is small). The cube is deformed to A'B'CD. If we now
imagine the cube rotated through 45° about the x_3 axis, it can be seen
that the same shear deformation could have been produced by the
equal tensile and compressive stresses as shown in Figure 1.21(b). The
stresses σ_{11} and σ_{-22} produce equal strains ϵ_{11} and ϵ_{-22}. Hence, from
equations 1.44 and remembering the simplified notation of writing a
single subscript in place of two alike subscripts and also that the stress
in the x_3 direction is zero, we may write

$$\epsilon_1 = -\epsilon_2$$

$$= \frac{1}{E}(\sigma_1 + \nu\sigma_1)$$

$$= \frac{(1+\nu)\,\sigma_1}{E} \qquad (1.53)$$

33

since $\sigma_1 = -\sigma_2$ and $\sigma_3 = 0$. By definition, the shear modulus (equation 1.5) is

$$G = \frac{\text{shear stress}, \tau}{\text{shear strain}, \gamma}$$

for small shears such that $\tan \gamma = \gamma$. But we have just shown that $\tau = \sigma_1$ and also, from equation 1.35, $\epsilon_1 = \frac{1}{2}\gamma$, so by substituting from equation 1.53

$$G = \frac{\sigma_1}{2\epsilon_1} = \frac{E}{2(1+\nu)} \tag{1.54}$$

By comparison with equation 1.48 it will be seen that we have proved the proposition that $G = G'$, as occurring in equation 1.49. Then, by combining equations 1.52 and 1.54,

$$E = \frac{9KG}{3K+G} \tag{1.55}$$

and

$$\nu = \frac{3K - 2G}{6K + 2G} \tag{1.56}$$

which relate Young's modulus E, the bulk modulus K, the shear modulus G, and Poisson's ratio ν.

Also, equation 1.56 may be rearranged so that

$$3K(1 - 2\nu) = 2G(1 + \nu) \tag{1.57}$$

Since K and G are both positive quantities, it follows that $1 - 2\nu$ and $1 + \nu$ must be of the same sign. This implies that ν must be less than $\frac{1}{2}$ and greater than -1, hence in theory $-1 < \nu < 0.5$; in practice, however, there is no known material that increases its transverse dimensions owing to a longitudinal tensile stress, so ν is in fact never negative.

Equation 1.55 relating the moduli may also be rearranged so that

$$\frac{1}{E} = \frac{1}{3G} + \frac{1}{9K} \tag{1.58}$$

For all materials, the bulk modulus K is greater than the shear modulus G, generally by a factor of about 3, thus $1/9K \ll 1/3G$. Young's modulus E therefore depends mainly on the value of the shear modulus, and very approximately

$$E \simeq 3G \tag{1.59}$$

How these moduli may be measured in practice is discussed in the next chapter.

1.6 Relationships between the Moduli – An Alternative Derivation

If we are concerned only with deriving relationships between the elastic moduli of an isotropic solid, then we can do so without direct reference to tensors and their corresponding notation. Although shorter, such an alternative derivation is, however, less satisfactory as it fails to give a deeper insight into the more general problems of elasticity.

By considering a unit cube of an isotropic material to be subjected to three stresses in directions parallel to the cube edges, we can conveniently define the stresses as σ_x, σ_y, and σ_z, which in turn produce corresponding strains ϵ_x, ϵ_y, and ϵ_z. To differentiate between tension and compression we can term a tensile stress as positive and a compressive stress as negative. By Hooke's law the strain produced will be proportional to the applied stress. However, the constant of proportionality may not be the same for both extension and contraction and therefore we must define separate constants. If s_E is the constant of proportionality between strain and the corresponding normal stress for an extension, and s_C likewise for a contraction, then the total extension of the unit cube in each of its three directions, remembering that an extension in one direction results in a contraction in the other two directions, is

$$\epsilon_x = s_E \sigma_x - s_C(\sigma_y + \sigma_z) \tag{1.60a}$$

$$\epsilon_y = s_E \sigma_y - s_C(\sigma_z + \sigma_x) \tag{1.60b}$$

$$\epsilon_z = s_E \sigma_z - s_C(\sigma_x + \sigma_y) \tag{1.60c}$$

To obtain a relationship between the moduli, we must find an expression for each of them in terms of s_E and s_C and then eliminate s_E and s_C between the expressions. For Young's modulus we are only concerned with an applied stress in one direction, say the x direction, so that $\sigma_y = \sigma_z = 0$, and the resulting strain in that same direction. Hence equations 1.60 simplify to

$$\epsilon_x = s_E \sigma_x \tag{1.61}$$

and, from equation 1.3,

$$\text{Young's modulus, } E = \frac{\sigma_x}{\epsilon_x} = \frac{1}{s_E} \tag{1.62}$$

Similarly, for the bulk modulus, since we must consider the stresses as being applied equally in the three directions to give equal linear

35

strains in the corresponding directions, that is, $\sigma_x = \sigma_y = \sigma_z$ and $\epsilon_x = \epsilon_y = \epsilon_z$, it follows from equations 1.60 that

$$\epsilon_x = (s_E - 2s_C)\,\sigma_x \qquad (1.63)$$

Also, the bulk or volume strain is three times the linear strain. This is easily seen since the edge dimensions of the unit cube have been changed from unity to $1 - \epsilon_x$ and thus its volume has been changed by an amount $1 - (1 - \epsilon_x)^3$. Hence the bulk strain, the change in volume per unit volume, is to a first approximation $3\epsilon_x$. The bulk modulus, the ratio of bulk stress to bulk strain (equation 1.4) is thus

$$\text{bulk modulus}, K = \frac{\sigma_x}{3\epsilon_x}$$

$$= \frac{1}{3(s_E - 2s_C)} \qquad (1.64)$$

from equation 1.63.

The shear modulus (equation 1.5), however, presents some problems since we have to relate the shear strain $\tan \gamma$, equal to γ for small strains, to the linear strain. That is, we have to obtain equation 1.35 but this time without the use of tensors. It is here that the alternative method of this section is at its least satisfactory. In Figure 1.22 a cube

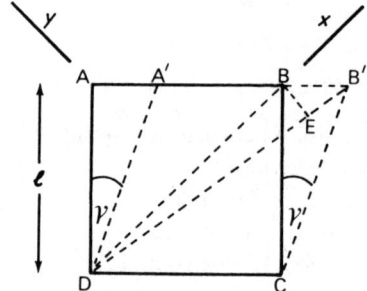

Figure 1.22. *Elevation cross-section of a cube of side l sheared through an angle* γ

of edge length l is shown sheared through a small angle γ. The diagonal DB has been increased to DB′ and the diagonal AC decreased to A′C. By making DE equal to DB we see that, since BB′ is small, angle BEB′ is approximately 90° and angle EB′B is approximately 45°, so that to a first approximation BE = EB'. Thus the strain along DB', the increase in length per unit length, is

$$\epsilon = \frac{DB' - DB}{DB}$$

$$= \frac{EB'}{DB}$$

$$= \frac{BB' \sin 45°}{l/\sin 45°}$$

$$= \frac{BB'}{\sqrt{2}} \frac{1}{l\sqrt{2}}$$

$$= \frac{1}{2} \frac{BB'}{l}$$

$$= \frac{1}{2} \gamma \qquad (1.65)$$

in agreement with the more rigorously derived equation 1.35.

We have also shown in the previous section, without the use of tensors, that an applied shearing stress is equivalent to an equal tensile stress along one diagonal and an equal compressive stress along the other. Thus the shear stress giving rise to the deformation shown in Figure 1.22 can be regarded as equal in magnitude to a tensile stress acting along DB (or DB' since BB' is small), and a compressive stress along AC (or A'C). If we regard these directions as the x and y directions respectively, then the shear stress $\tau = \sigma_x = -\sigma_y$, and $\sigma_z = 0$. The shear modulus may then be written as

$$\text{shear modulus, } G = \frac{\tau}{\gamma} = \frac{\sigma_x}{2\epsilon_x} \qquad (1.66)$$

from equation 1.65. Also, from equations 1.60,

$$\epsilon_x = (s_E + s_C) \sigma_x \qquad (1.67)$$

and hence

$$G = \frac{1}{2(s_E + s_C)} \qquad (1.68)$$

If we now eliminate s_E and s_C between equations 1.62, 1.64, and 1.68, for example by substituting s_C from equation 1.68 into equation 1.64 and then substituting for s_E from equation 1.62, we obtain

$$E = \frac{9KG}{3K + G} \tag{1.69}$$

in agreement with equation 1.55.

Finally, we have to derive a connection with Poisson's ratio. This follows quite simply from the definition (equation 1.6) since

$$\text{Poisson's ratio, } \nu = -\frac{\epsilon_y}{\epsilon_x} \tag{1.70}$$

for $\sigma_y = \sigma_z = 0$. Substituting from equations 1.60,

$$\nu = +\frac{s_C \sigma_x}{s_E \sigma_x} = \frac{s_C}{s_E} \tag{1.71}$$

By obtaining expressions for s_C and s_E from equations 1.64 and 1.68 and substituting in equation 1.71,

$$\nu = \frac{3K - 2G}{6K + 2G} \tag{1.72}$$

in agreement with equation 1.56.

Alternatively, by assuming the final result, equation 1.71 may be quickly verified by substituting the values of K and G from equations 1.64 and 1.68 directly into equation 1.72. The deductions made at the end of Section 1.5 then follow as before.

1.7 Strain Energy

If a body is to be strained, then the forces producing that strain must do work. Once the body is strained it may itself be considered as capable of doing work, since in the strained state it is exerting forces on the surrounding objects that are maintaining the strain. The forces that the body is exerting on its surroundings are exactly equal to the forces that must be exerted by the surroundings to produce that strain, providing that the body is perfectly elastic. If now the strained body is allowed to recover slowly its original form, the forces exerted at every instant by the body on its surroundings will be equal to those provided by the surroundings in producing that strain in the first place. That is, as the body returns to its unstrained condition it gives back all the

38

work put into it to produce the original strain. Thus a strained body may be considered as possessing potential energy, termed the *strain energy*, equal to the external work done in producing the strain.

Let us consider a unit cube with edges parallel to the x_1, x_2, and x_3 directions and with the centre of coordinates at a corner of the cube as shown in Figure 1.23. A stress σ_{11} produces a strain ϵ_{11}. Let now

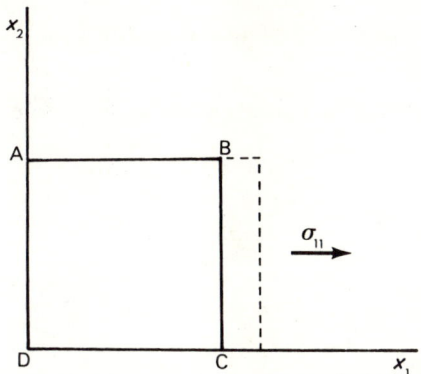

Figure 1.23. Application of a stress σ_{11} to a unit cube

this strain component be increased to $\epsilon_{11} + \delta\epsilon_{11}$. This means that the face BC normal to the x_1 axis will move through an extra distance $\delta\epsilon_{11}$. The work done on this face is equal to the normal component of force multiplied by the displacement, that is, $\sigma_{11} \delta\epsilon_{11}$. Similar expressions pertain for stresses σ_{22} and σ_{33}. If now the cube is sheared by applying a stress σ_{12}, causing the face BC to move in the x_2 direction, a shear strain ϵ_{12} will be produced. Again let this strain be increased by a small amount to $\epsilon_{12} + \delta\epsilon_{12}$. The work done is therefore $\sigma_{12} \delta\epsilon_{12}$. Again there are similar expressions for the other shear stresses σ_{23} and σ_{31}, remembering also that $\sigma_{ij} = \sigma_{ji}$ (equation 1.19) and $\epsilon_{ij} = \epsilon_{ji}$. Thus generally the strain energy, equal to the work done by stress components σ_{ij}, is given by

$$dW = \sigma_{ij}\, d\epsilon_{ij} \qquad (1.73)$$

where i and j = 1, 2, 3, for small increments of strain. We may illustrate how to calculate the strain energy by considering two simple examples.

Consider a piece of wire of length l and cross-sectional area S. An axial stress σ applied to the wire will produce a strain ϵ, so the length increases by $l\epsilon$. If now the length of the wire is increased further by a small amount $\delta(l\epsilon)$, the extra work required to provide this increment of strain is

$$\delta W = \text{force} \times \text{distance}$$

$$= (\sigma S)\,\sigma(l\epsilon)$$

Therefore the total work required to produce the strain ϵ is

$$W = \int_0^\epsilon \sigma S l\,d\epsilon$$

But $\sigma = \epsilon E$, where E is Young's modulus. Therefore

$$W = S l E \int_0^\epsilon \epsilon\,d\epsilon$$

$$= \frac{1}{2} S l E\,\epsilon^2$$

$$= \frac{1}{2} \sigma \epsilon (S l)$$

That is, the strain energy, the work necessary to produce a strain ϵ by applying a corresponding stress σ, is

$$W = \frac{1}{2}\sigma\epsilon \qquad (1.74)$$

per unit volume, since $S l$ is the volume of the wire.

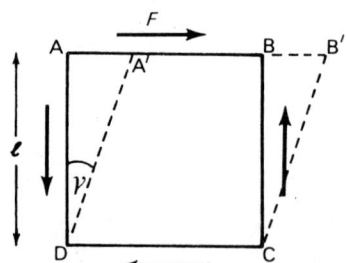

Figure 1.24. Production of a shear strain γ by the application of a force F on a cube

Similarly the strain energy may be calculated for a body subjected to a shear stress as follows. A shear strain $\tan \gamma = \gamma$ is produced by the application of a force F on a cube of face area l^2, as shown in Figure 1.24. The work involved in producing this shear strain is

40

$$W = \text{mean force} \times \text{distance}$$

$$= \frac{1}{2}F \; (\text{AA}')$$

$$= \frac{1}{2}l^2 \, \tau \, (l\gamma)$$

$$= \frac{1}{2}\tau\gamma(l^3)$$

where τ is the shear stress. That is, the strain energy, the work required to produce a shear strain γ with a shear stress τ, is

$$W = \frac{1}{2}\tau\gamma \qquad (1.75)$$

per unit volume.

1.8 Resilience

Resilience, or restitution, is the ability of a strained body to return to its original unstrained form. It is a measure of its ability to resist a mechanical blow and not acquire any permanent deformation. It is measured by the amount of work required to strain a body without exceeding the elastic limit and is thus numerically equal to the strain energy.

For example, consider the straining of a piece of wire, discussed in the previous section. We showed that the strain energy per unit volume was $W = \frac{1}{2} \sigma\epsilon$ (equation 1.74). Hence the resilience per unit volume is

$$W = \frac{\sigma^2}{2E} \qquad (1.76)$$

where Young's modulus $E = \sigma/\epsilon$. If σ_0 is the maximum stress that can be applied to this piece of wire without exceeding the elastic limit, then the maximum strain energy that can be stored in the wire without causing a permanent deformation is

$$W_0 = \frac{\sigma_0^2}{2E} \qquad (1.77)$$

where W_0 is termed the *proof resilience*.

Chapter Two

APPLICATIONS OF ELASTICITY

2.1 Introduction

In this chapter we shall consider the more practical applications of elasticity. The previous chapter dealt with the theoretical concepts of what is meant by elasticity and stress and strain and how the various moduli are interrelated. Measurements of the moduli enable the shape and size of a particular body after it has been stressed to be predicted, but only if the stress is applied in the same way as prescribed for that particular modulus. What happens if a rod is twisted or a beam is bent? Measurements of the same moduli are still available, but the problem must be adapted so as to make their application relevant. Having discussed the connections between the deformation of various shaped bodies and various moduli, we may return to a discussion of how the moduli are determined. It is not always best to deform a body in a manner as directly indicated in the definitions, but may be better to deform some other shaped body and determine the modulus indirectly. We shall then go on to consider the general measurement of strain in practice as produced by a given stress, and finally we shall discuss the relevance of elasticity in liquids, for even a liquid has a tensile strength.

2.2 Torsion in Rods

Let us consider now the effect of applying a twisting couple to a wire or rod. We wish to know how much it will twist and how the twist is related to a measurable quantity for the material, in this case the rigidity modulus.

Consider a rod or wire of length l. In this, imagine an elemental coaxial cylindrical tube of internal radius r, thickness δr, and length l. Suppose now that a torque is applied to the rod, as shown in Figure 2.1. One end of the rod will then twist through an angle θ relative to the other, so an element of area $\delta r \, \delta s$ on the elemental annular end-face of the tube will move from B to C, as in the figure. The tube has now been sheared by an angle γ. Providing both these angles are small,

$$l\gamma = r\theta \qquad (2.1)$$

If τ is the tangential stress on the elemental tube, that is, the tangential force per unit area acting on its annular end-face, due to a couple T

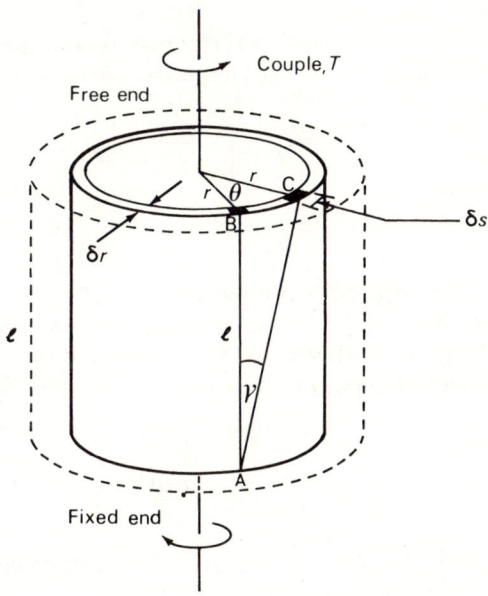

Figure 2.1.   *Application of a twisting couple to a rod*

applied to the ends of the whole rod, then the tangential force acting on the elemental area $\delta r\ \delta s$ is $\tau\ \delta r\ \delta s$. That is, the element of couple about the rod axis contributed by this force is

$$\delta T = (\tau\ \delta r\ \delta s)r$$
$$= \frac{\tau\theta\,(\delta r\ \delta s)r^2}{l\gamma} \qquad (2.2)$$

from equation 2.1. Also, from the definition of the shear modulus (equation 1.5), $G = \tau/\gamma$ for small shear, so

$$\delta T = \frac{G\theta\,(\delta r\ \delta s)r^2}{l} \qquad (2.3)$$

43

Therefore the total couple on the complete annular end-face of the tube is

$$\delta T = \frac{G\theta}{l} \Sigma (\delta r \ \delta s) r^2$$

$$= \frac{2\pi G\theta}{l} r^3 \ \delta r \qquad (2.4)$$

since $\Sigma \delta s = 2\pi r$

Hence the total couple that is required to twist the whole rod of radius r will be the sum of the elemental annular couples, that is, the total couple is

$$T = \int_0^r \frac{2\pi G\theta}{l} r^3 \ dr$$

$$= \frac{\pi G r^4}{2l} \theta \qquad (2.5)$$

where θ is the angle of twist produced in a rod of length l.

We can also calculate the strain energy for a twisted rod (see Section 1.7). The work required to twist the rod through a small angle $\delta\theta$ by a couple T applied to the rod is

$$\delta W = T \ \delta\theta$$

$$= \frac{\pi G r^4}{2l} \theta \ \delta\theta \qquad (2.6)$$

from equation 2.5. Hence the total work (the strain energy) involved in twisting the rod through an angle θ is

$$W = \frac{\pi G r^4}{2l} \int_0^\theta \theta \ d\theta$$

$$= \frac{\pi G r^4}{2l} \frac{\theta^2}{2}$$

$$= \frac{T\theta}{2} \qquad (2.7)$$

or

$$W = \frac{T^2 l}{\pi G r^4} \qquad (2.8)$$

by substituting from equation 2.5. We also see from equation 2.7 that the strain energy (the potential energy) of the twisted rod can be written as $\frac{1}{2}k\theta^2$, where k is the torsional constant of the rod, equal to $\pi G r^4 / 2l$.

44

2.3 Bending of Beams

If a horizontal beam is loaded at some point, then there will obviously be some deflection. We can relate the deflection to the load without much difficulty providing we restrict the problem to one of simple cross-sections for the beam and deflections into circular arcs, and work within the elastic limit of the beam material. The method is to relate the bending moment of the beam with Young's modulus and with the deflection. The general expressions are then applied to the particular problem involving a given shape of beam, method of support, and its method of loading.

Let us consider first a beam bent into a circular arc of radius r, as shown in Figure 2.2(a). The beam material on the convex side of the

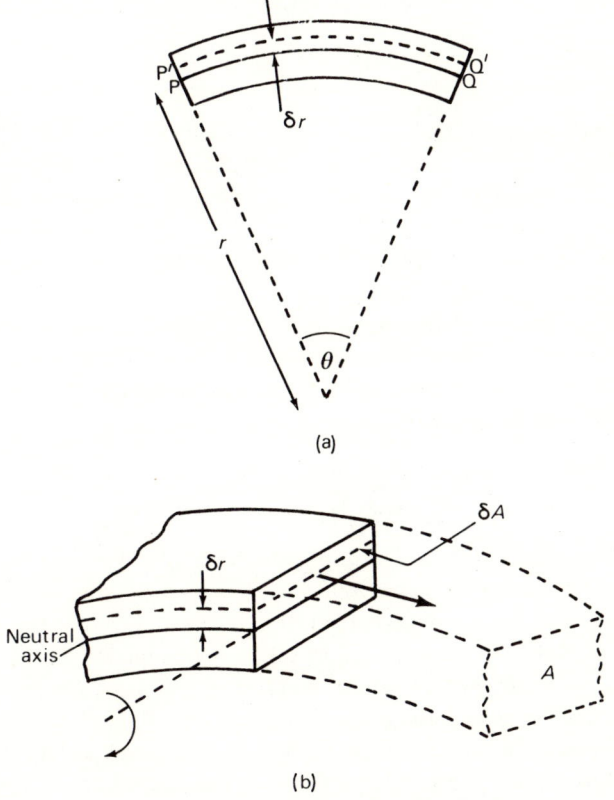

(a)

(b)

Figure 2.2.   Bending of a beam

curve will be stretched, whilst that on the concave side will be compressed. Between the two will occur an element PQ which has not changed its length. This is termed the *neutral axis*. We may also consider another element P'Q' parallel to the neutral axis and distance δr from it so that

$$\theta = \frac{PQ}{r} = \frac{P'Q'}{r + \delta r} \tag{2.9}$$

The linear strain of the element P'Q' is then

$$\epsilon = \frac{P'Q' - PQ}{PQ}$$

$$= \frac{(r + \delta r)\,\theta - r\theta}{r\theta}$$

$$= \frac{\delta r}{r} \tag{2.10}$$

If the area of the normal cross-section of the beam is A and of the element is δA, then the longitudinal stress on this element is $\sigma = \epsilon E$, where E is Young's modulus; the longitudinal force being consequently $\epsilon E \delta A = E\,\delta A\,\delta r/r$.

This longitudinal force exerts a moment about a transverse axis passing through, and normal to, the neutral axis as shown in Figure 2.2(b), the moment being equal to $E\delta A (\delta r)^2 /r$. The sum of all such moments of force about this axis will constitute a couple tending to bend the beam into an arc, and will be equal to the couple that is applied to the beam, that is,

$$T = \frac{E}{r}\; \Sigma \delta A (\delta r)^2 \tag{2.11}$$

The term $\Sigma \delta A (\delta r)^2$ may be compared with the expression for the moment of inertia $I = \Sigma \delta m (\delta r)^2 = mk^2$, where m is the total mass of the body and k is the radius of gyration of the body about the axis of rotation $\delta r = 0$. By analogy $\Sigma \delta A (\delta r)^2$ may be replaced by Ak^2 where A is the total cross-sectional area and k is the radius of gyration as before. Ak^2 is termed the *geometrical moment of inertia* or the *second areal moment* of the area. Figure 2.3 shows values for Ak^2 for horizontal axes through the 'centre of gravity' of various shaped beam cross-sections.

46

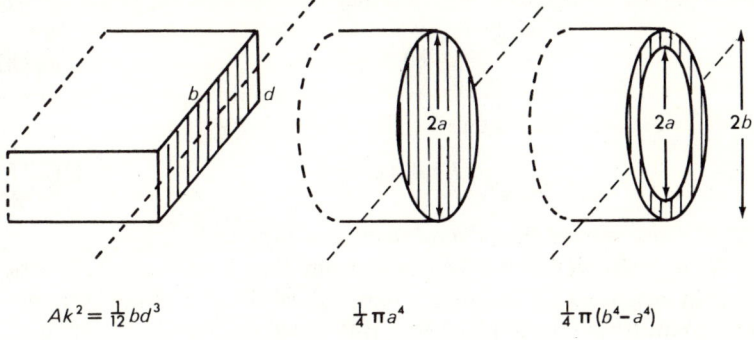

$Ak^2 = \frac{1}{12}bd^3$ $\frac{1}{4}\pi a^4$ $\frac{1}{4}\pi(b^4-a^4)$

Figure 2.3.   *Geometrical moments of inertia for beams of different cross-sections*

Then the applied couple T is given by

$$T = \frac{EAk^2}{r} \tag{2.12}$$

Although this is an exact expression and may be generally applied even to beams with considerable curvature, a more convenient working approximation may be made for beams involving only small deflections

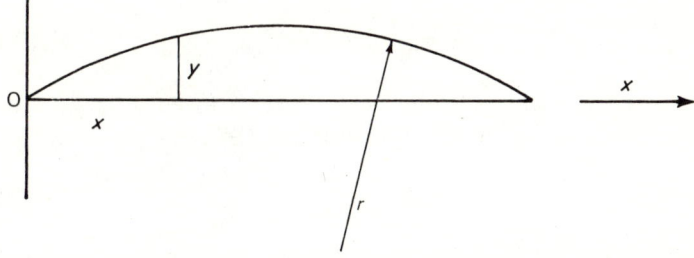

Figure 2.4.   *Transposition of radius of curvature r into xy coordinates*

from the straight. The deflection of the beam in terms of the radius of curvature r may be transposed into more conveniently measurable xy coordinates since from coordinate geometry (see Figure 2.4)

$$\pm\frac{1}{r} = \frac{d^2y/dx^2}{[1 + (dy/dx)^2]^{3/2}} \tag{2.13}$$

47

Since in practice $(dy/dx)^2 \ll 1$, this may be simplified to

$$\pm\frac{1}{r} = \frac{d^2y}{dx^2} \tag{2.14}$$

so that equation 2.12 becomes

$$T = EAk^2 \frac{d^2y}{dx^2} \tag{2.15}$$

EAk^2 is also termed the *flexural rigidity* of the beam.

We may also determine the strain energy for a bent beam. The work done in increasing the length of a longitudinal element of the beam by an amount from zero to ϵl, where ϵ is the final strain and l the original length, is

$$\delta W = \text{mean force} \times \text{distance}$$

$$= \left(\frac{1}{2}\sigma\,\delta A\right)(\epsilon l)$$

where σ is the stress producing a strain ϵ and acts over the cross-sectional area δA of the element. Also $\sigma = \epsilon E$, where E is Young's modulus and $\epsilon = \delta r/r$ by equation 2.10, so

$$\delta W = \frac{El}{2r^2}\,\delta A\,(\delta r)^2 \tag{2.16}$$

Hence the total strain energy of the beam will be

$$W = \frac{El}{2r^2}\,\Sigma\delta A\,(\delta r)^2 \tag{2.17}$$

But $\Sigma\delta A(\delta r)^2 = Ak^2$ as before and $r = EAk^2/T$ by equation 2.12, hence the total strain energy of the beam of length l is

$$W = \frac{T^2 l}{2EAk^2} \tag{2.18}$$

The solution of a particular beam problem generally becomes one of determining an expression for the applied couple T and applying boundary conditions in order to solve the second-order differential equation (equation 2.15) so as to determine the deflection y at a distance x from some convenient origin. To illustrate the method we shall examine a number of different types of beam problem.

The simplest problem to consider is that of a light horizontal cantilever projecting from a vertical wall and supporting a mass m at the other end, as shown in Figure 2.5(a). There will be a vertical reaction R equal and opposite to the applied force mg. These two forces constitute a shearing action which is negligible in comparison with the deflection due to the bending moment mgl. There will also be an equal and opposite couple exerted by the wall. The beam being light, its weight may be neglected in comparison with the applied load.

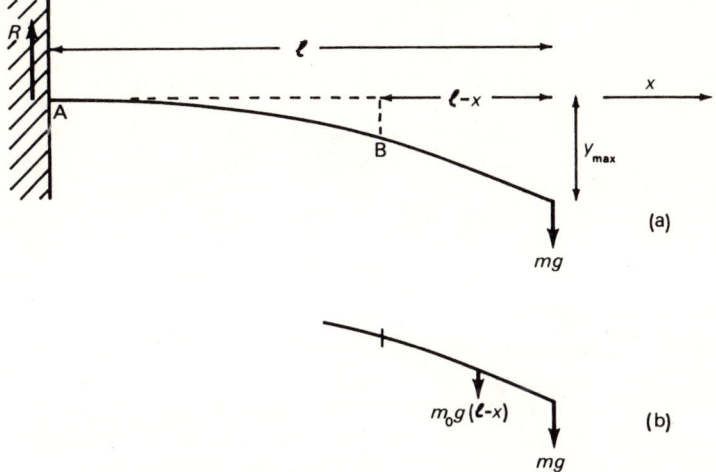

Figure 2.5.   *Bending of a horizontal cantilever: (a) a light beam, and (b) a beam of mass m_0 per unit length*

At some point B, distance x from the origin which is here taken as the point where the cantilever joins the supporting wall, the couple acting to deflect the beam will be

$$T = mg(l - x)$$

Then, from equation 2.15,

$$mg(l - x) = EAk^2 \frac{d^2 y}{dx^2} \qquad (2.19)$$

This is a second-order equation; therefore, integrating twice,

$$mglx - \frac{mgx^2}{2} + B = EAk^2 \frac{dy}{dx} \qquad (2.20)$$

and

$$\frac{mglx^2}{2} - \frac{mgx^3}{6} + Bx + C = EAk^2 y \qquad (2.21)$$

49

where B and C are constants of integration and depend on the problem. In this particular case there is no depression at the origin, that is, $y = 0$ when $x = 0$. Therefore $C = 0$ from equation 2.21. Also, at the origin the cantilever is horizontal, so $dy/dx = 0$ when $x = 0$. Thus $B = 0$ from equation 2.20 and

$$\frac{mglx^2}{2} - \frac{mgx^3}{6} = EAk^2 y \qquad (2.22)$$

The depression y at a point distance x from the origin is then

$$y = \frac{mgx^2 (3l - x)}{6EAk^2} \qquad (2.23)$$

The maximum depression y_{max} occurs at $x = l$, that is,

$$y_{max} = \frac{mgl^3}{3EAk^2} \qquad (2.24)$$

where Ak^2 will depend on the cross-sectional shape of the particular beam.

If, however, the cantilever is not light but has a mass m_0 per unit length, then this will increase the deflection. The turning couple at point B is now increased by

$$m_0 g(l - x)\left[\frac{1}{2}(l - x)\right] = \frac{1}{2} m_0 g(l - x)^2$$

as in Figure 2.5(b), so the total turning moment is

$$mg(l - x) + \frac{1}{2} m_0 g(l - x)^2 = EAk^2 \frac{d^2 y}{dx^2} \qquad (2.25)$$

Again this may be integrated twice, and with the same boundary conditions as before the constants of integration are again zero. Thus the deflection y_{max} at the end of the beam, when $x = l$, is

$$y_{max} = \frac{1}{EAk^2}\left(\frac{mgl^3}{3} + \frac{m_0 gl^4}{8}\right) \qquad (2.26)$$

In particular, if the total weight of the beam is $Mg = m_0 gl$, then

$$y_{max} = \frac{mgl^3}{3EAk^2}\left(1 + \frac{3M}{8m}\right) \qquad (2.27)$$

which may be compared with equation 2.24 for a light beam.

50

If now a light horizontal beam is supported freely at its ends and carries a central load, as shown in Figure 2.6, then the same procedure may be followed. However, the reactions at the supports must be taken into account as it is here that the bending couple is being exerted,

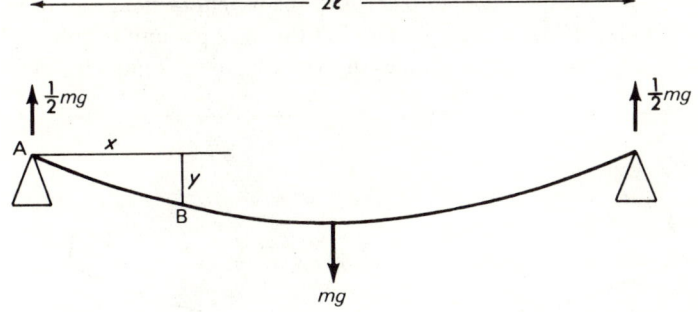

Figure 2.6.   *Bending of a light horizontal beam carrying a central load*

not at the point of maximum deflection as before. As in the previous examples we consider the couple at some point B, distance x $(0 \leqslant x \leqslant l)$ from the origin at A. The reaction $\frac{1}{2}\,mg$ is acting vertically upwards, so the bending moment about B is $-\frac{1}{2}\,mgx$. Hence

$$-\frac{1}{2}\,mgx = EAk^2\,\frac{\mathrm{d}^2 y}{\mathrm{d}x^2} \tag{2.28}$$

As before, integrating this twice gives at the first integration

$$\frac{\mathrm{d}y}{\mathrm{d}x} = -\frac{mgx^2}{4EAk^2} + B \tag{2.29}$$

Applying the boundary condition that $\mathrm{d}y/\mathrm{d}x = 0$ at $x = l$ gives $B = mgl^2/4EAk^2$ and

$$\frac{\mathrm{d}y}{\mathrm{d}x} = \frac{mg}{EAk^2}\left(\frac{l^2}{4} - \frac{x^2}{4}\right) \tag{2.30}$$

The second integration, with the boundary condition that $y = 0$ at $x = 0$, gives

$$y = \frac{mg}{EAk^2}\left(\frac{l^2 x}{4} - \frac{x^3}{12}\right) \tag{2.31}$$

for the deflection y at a point x from the end of the beam. In particular, the maximum deflection y_{max} occurring at the centre of the beam, where $x = l$, is

$$y_{max} = \frac{mgl^3}{6EAk^2} \qquad (2.32)$$

If it is necessary to take into account the weight of the beam itself then this may be done quite simply. Let the mass per unit length of the beam be m_0, then the extra forces shown in Figure 2.7 must be taken

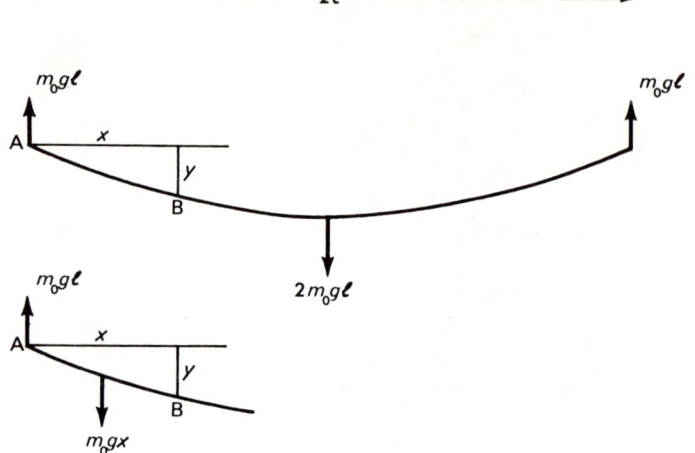

Figure 2.7.   Bending of a beam, taking into account its weight

into account. The extra bending moment about B is thus $\frac{1}{2} m_0 g x^2 - m_0 g l x$. If there is a central load, as before the total bending moment about B will be

$$\frac{1}{2} m_0 g x^2 - m_0 g l x - \frac{1}{2} mgx = EAk^2 \frac{d^2 y}{dx^2} \qquad (2.33)$$

The deflection y may be determined by integrating and applying the boundary conditions $dy/dx = 0$ at $x = l$ and $y = 0$ at $x = 0$.

The solution of the beam problems so far described have all involved second-order differential equations, and the problems have resolved themselves into deciding the bending moments and the two constants of integration. This approach is quite adequate to deal with beams having only two simple supports. For more complicated cases, for example a built-in or encastré beam in which the beam is loaded in the

52

middle but the two ends are constrained to be horizontal, as shown in Figure 2.9, then it is not possible to calculate the bending moment as we have done as it is necessary to bring in the other forces that are constraining the ends. We may extend our basic equation (equation 2.15) in the following way.

Consider a section of length δx measured along the neutral axis of a beam, as shown in Figure 2.8. The shearing forces F and $F + \delta F$ and

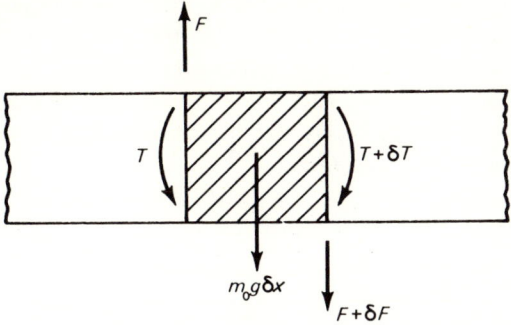

Figure 2.8.   Forces acting on a section of a beam

the bending moments T and $T + \delta T$ will act on the section as shown, as well as $m_0 g \delta x$ where m_0 is the mass per unit length of the beam. For equilibrium of this section,

$$(F + \delta F) + m_0 g \delta x = F$$

so

$$\frac{dF}{dx} = -m_0 g \qquad (2.34)$$

and, by taking moments about the left-hand end,

$$(T + \delta T) + m_0 g \, \delta x \left(\frac{1}{2} \delta x\right) + (F + \delta F) \, \delta x = T$$

so

$$\frac{dT}{dx} = -F \qquad (2.35)$$

by neglecting the products of small quantities. From equations 2.34 and 2.35,

$$\frac{d^2 T}{dx^2} = m_0 g \qquad (2.36)$$

53

Combining this with equation 2.15 gives

$$\frac{d^2 T}{dx^2} = EAk^2 \frac{d^4 y}{dx^4} = m_0 g \qquad (2.37)$$

The use of this equation involves four constants of integration which for most beam problems are sufficient to incorporate the extra boundary conditions arising from the method of support. This equation could as well have been used in the previous problem, for example, but it would have been necessary to make use of the fact that the bending moment is zero at the freely supported end of a beam. The extra boundary conditions would have come from the fact that the bending moment, and consequently $d^2 y/dx^2$ by equation 2.15, is zero at $x = 0$ and also $y = 0$ at $x = 2l$.

As an example of the use of equation 2.34 we shall consider the problem of the built-in beam illustrated in Figure 2.9. The ends of the

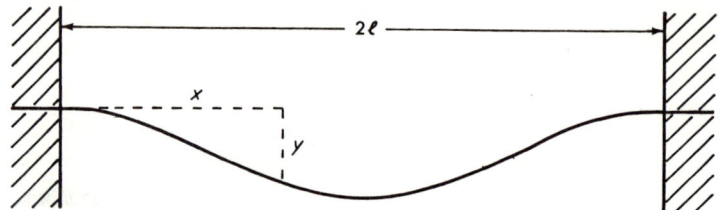

Figure 2.9. *Bending of a built-in beam*

beam of length $2l$ are held in a horizontal position and in line. Let the mass per unit length of the beam be m_0 and let it be sagging under its own weight. Equation 2.37 must then be integrated four times, giving:

$$EAk^2 \frac{d^4 y}{dx^4} = m_0 g$$

$$EAk^2 \frac{d^3 y}{dx^3} = m_0 g x + B \qquad (2.38)$$

$$EAk^2 \frac{d^2 y}{dx^2} = \frac{m_0 g x^2}{2} + Bx + C \qquad (2.39)$$

$$EAk^2 \frac{dy}{dx} = \frac{m_0 g x^3}{6} + \frac{Bx^2}{2} + Cx + D \qquad (2.40)$$

$$EAk^2 y = \frac{m_0 g x^4}{24} + \frac{Bx^3}{6} + \frac{Cx^2}{2} + Dx + F \qquad (2.41)$$

54

At the wall (that is, at $x = 0$) $dy/dx = y = 0$. Therefore, from equations 2.40 and 2.41, $D = F = 0$. Also $dy/dx = y = 0$ again at the other wall, at $x = 2l$, so equations 2.40 and 2.41 become

$$\frac{4}{3} m_0 g l^3 + 2Bl^2 + 2Cl = 0 \tag{2.42}$$

and

$$\frac{2}{3} m_0 g l^4 + \frac{4}{3} Bl^3 + 2Cl^2 = 0 \tag{2.43}$$

from which

$$B = -m_0 g l \tag{2.44}$$

and

$$C = \frac{1}{3} m_0 g l^2 \tag{2.45}$$

Therefore equation 2.41 becomes

$$EAk^2 y = \frac{m_0 g x^4}{24} - \frac{m_0 g l x^3}{6} + \frac{m_0 g l^2 x^2}{6} \tag{2.46}$$

Thus the deflection y_{max} at the middle of the beam, where $x = l$, is

$$y_{max} = \frac{m_0 g l^4}{24 EA k^2} \tag{2.47}$$

Equation 2.39 also gives the bending moment at any point x along the beam since, by substituting from equations 2.44 and 2.45, the bending moment

$$T = EAk^2 \frac{d^2 y}{dx^2} = \frac{m_0 g x^2}{2} - m_0 g l x + \frac{m_0 g l^2}{3} \tag{2.48}$$

This will be zero when $\frac{1}{2} x^2 - lx + \frac{1}{3} l^2 = 0$ and therefore, solving the quadratic, this will be at the points given by $x = l \pm l/\sqrt{3}$.

2.4 Bending of Rods under Longitudinal Compression

A problem somewhat similar to the bending of a beam is that of the bending of a rod whilst being compressed longitudinally, for example a vertical rod supporting a load as shown in Figure 2.10, where the load is assumed capable of moving only along a vertical axis. The rod is assumed to be thin compared with its length, and the ends are rounded.

For small compressive loads, the rod will remain straight. A lateral force F will cause the rod to bend, but on its removal the rod will straighten again. If the compressive load is steadily increased, a time will come when the rod will not straighten on removal of the lateral

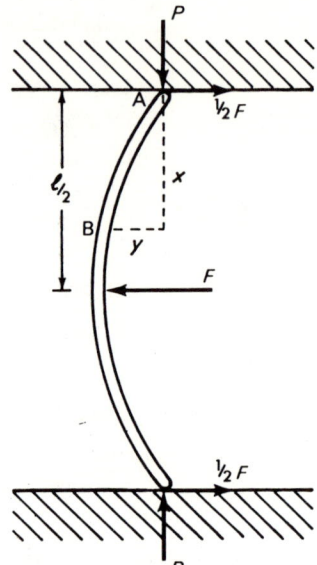

Figure 2.10. Bending of a rod under longitudinal compression

force and a *critical load* has been reached. Further increase in the compressive load leads to increased buckling of the rod and its ultimate collapse. To determine this critical load we may consider the bending moment on the rod and treat it in the same way as we did the bending of a beam.

Let us consider a point B on the rod, distance x from one end and with a deflection y from the straight condition. For equilibrium, the shearing force F will be balanced by two forces $\frac{1}{2}F$, as shown in Figure 2.10. Then, by taking moments about point B, we see that the condition for equilibrium of the upper half of the rod is that

$$\frac{1}{2}Fx + Py + F\left(\frac{1}{2}l - x\right) + T = 0 \tag{2.49}$$

where P is the compressive force causing the buckling, l is the length of the rod, and T is the bending moment on the rod. At the onset of buckling, when P is equal to the critical load, the shearing force F will also be zero so, for the critical load condition, equation 2.49 becomes

$$Py + T = 0 \qquad (2.50)$$

But the bending moment T is given by equation 2.15, and therefore

$$Py + EAk^2 \frac{d^2y}{dx^2} = 0$$

that is,

$$\frac{d^2y}{dx^2} + m^2y = 0 \qquad (2.51)$$

where $m^2 = P/EAk^2$. This equation is of the same form as that for simple harmonic motion where it may be recalled that the solution is

$$y = a \sin(mx - \alpha) \qquad (2.52)$$

where a and α are constants depending on the particular problem. That is, the rod buckles into a sine-wave form.

In our problem, $y = 0$ at $x = 0$. Therefore

$$0 = -a \sin \alpha$$

Hence $a = 0$ or $\alpha = 0, \pi, 2\pi, \ldots$. The value $a = 0$ would imply, by equation 2.52, that y is always zero and the rod straight, and so this value may be disregarded if the critical load P is taken as one that just begins to produce some buckling of the rod. We have also that $y = 0$ again at $x = l$, and therefore

$$0 = a \sin(ml - \alpha) \qquad (2.53)$$

Hence

$$ml = 0, \pi, 2\pi, \ldots$$

since $\alpha = 0, \pi, 2\pi, \ldots$ and $a \neq 0$. Also, since $m^2 = P/EAk^2$, the condition $ml = 0$ may be disregarded as it would imply that m and consequently P were zero and the rod would then be straight. Therefore

$$ml = \frac{l}{k}\sqrt{\frac{P}{EA}} = \pi, 2\pi, \ldots$$

that is, the critical load

$$P = \frac{n^2\pi^2 EAk^2}{l^2} \qquad (2.54)$$

57

where $n = 1, 2, \ldots$. The rod therefore buckles into the shape of a half-sine wave for $n = 1$ and into multiples of a half-sine wave for $n = 2, 3, \ldots$ for a critical load P which is just sufficient to keep the rod bent.

If a tapered rod or a rod in which the loading is not axial is considered, the problem becomes very complex. Even the result of the simple case considered above agrees only approximately with experiment, although it is possible to make a far more detailed analysis of the problem with slightly better agreement with experiment.

2.5 Helical Springs

Another problem that is closely related to the bending of a beam and also torsion in rods is the stretching of a helical spring. Here the wire of the spring is subjected to both a torque and a bending moment, which together produce a tendency generally for the spring to coil up when stretched.

The simplest treatment involves a tightly coiled spring, that is, one in which the individual coils may be regarded as each being in one plane. In an approximate treatment it is possible to neglect the bending moment for small extensions.

Consider the spring shown in Figure 2.11 which is being stretched by an axial force P and has a mean diameter $2r$. In particular, let us consider a short section, length l, of one of the coils. The torque on this section will be Pr, given (from equation 2.5) by

$$Pr = \frac{\pi G a^4}{2l} \theta \qquad (2.55)$$

where $2a$ is the diameter of the wire and θ is the angle of twist between the two ends of the section of coil. G is the shear modulus of the wire material. This twist causes the applied force P to move through a distance $r\theta$, which is the same as saying that this part of the spring stretches axially an amount $r\theta$ under a force P.

For n complete coils of the spring, each coil being of length $2\pi r$, the total length of wire subject to torsion will be $l = 2n\pi r$, so the total angle of twist of the full length of wire will be

$$\theta = \frac{4nr^2}{Ga^4} P \qquad (2.56)$$

from equation 2.55. Hence the elongation of the complete spring under an axial force P will be $z = r\theta$, that is,

$$z = \frac{4nr^3}{Ga^4} P \qquad (2.57)$$

When the spring is not closely coiled or is subjected to greater elongation, it becomes necessary to take account of the bending moment on

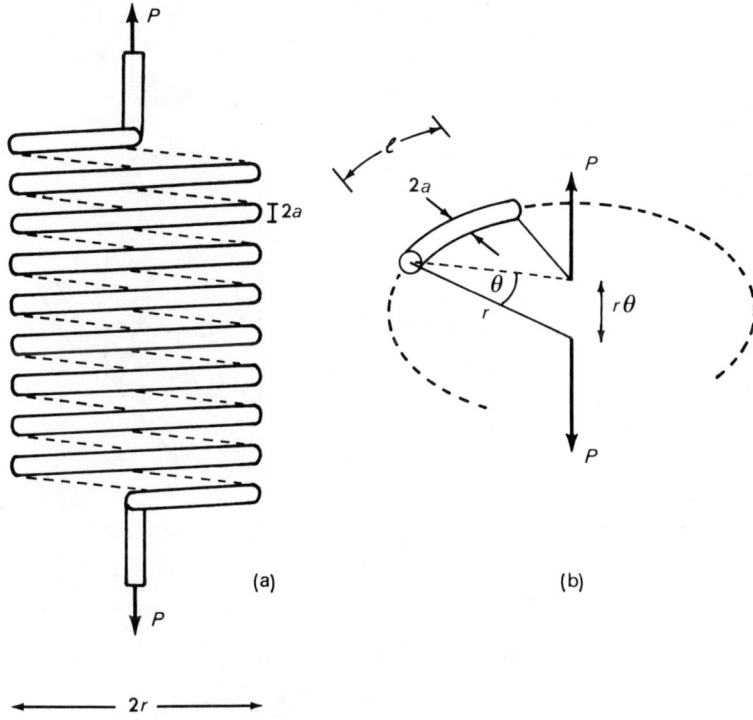

(a)

(b)

Figure 2.11. Stretching of a tightly coiled spring

the spring wire. Consider the open coiled spring of Figure 2.12(a) of mean diameter $2r$ and where the coils are inclined at an angle 2α to each other, as in the figure. Let the spring be stretched by an axial force P as before. Now the axis of the applied torque is no longer along the axis of the wire since the torque vector will be in a plane perpendicular to the direction of the spring axis, as shown in Figure 2.12(b). Thus, if the spring axis is vertical, the radius vector r and the torque vector T will both lie in a horizontal plane.

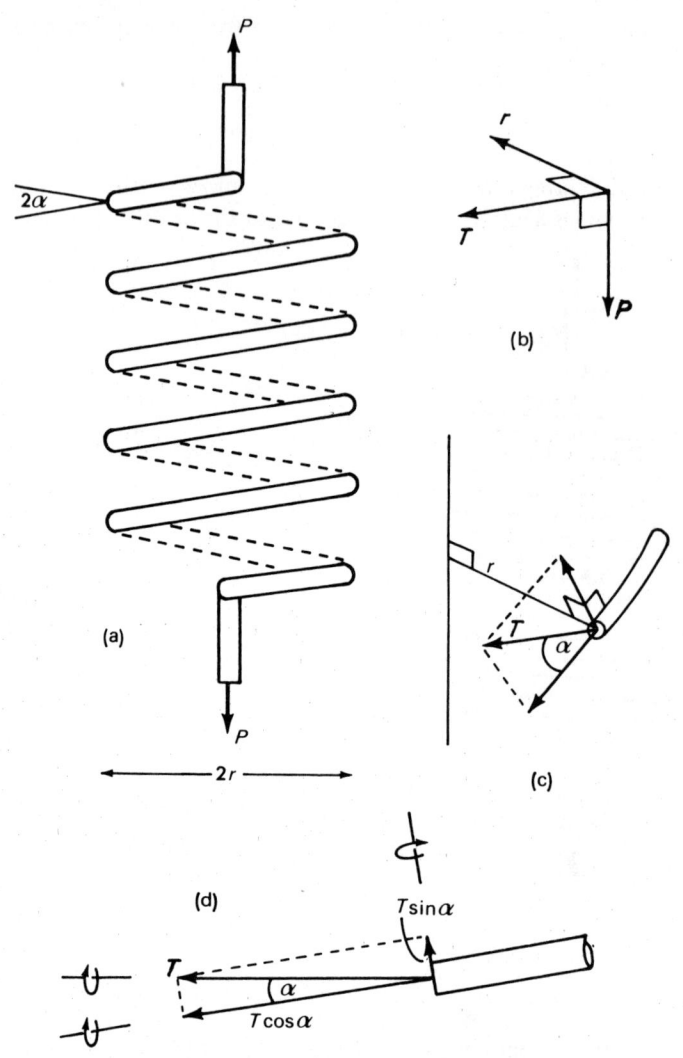

Figure 2.12. Stretching of an open coiled spring

From the definition of a vector product, the direction of the vector **r** must be rotated through an angle θ to make it parallel to the direction of the vector **P**, so the product of these two vectors will be another vector of magnitude $rP \sin \theta$ acting at right angles to the plane **rP** and in a direction given by an advancing right-hand screw motion due to a rotation of θ of $\mathbf{r}$ to $\mathbf{P}$. In this case, $\sin \theta$ is unity and therefore we may write $\mathbf{T} = \mathbf{r} \times \mathbf{P} = rP$.

Therefore $\mathbf{T}$ may be resolved into two components, one along the axis of the wire which will produce a twist in the wire and the other at right angles to the wire and in a plane parallel to the spring axis which will produce a bending moment tending to change the diameter of the helix. The directions of these components are illustrated in Figure 2.12(c) where a small section of a coil is considered. The plane of the coil is at an angle α to the horizontal. Figure 2.12(d) shows another view of this section of coil, this time looking in a horizontal direction towards the spring axis. Thus the torque on the wire is $T_1 = rP \cos \alpha$ and the bending moment $T_2 = rP \sin \alpha$.

Now the total work done in stretching the spring by an amount z from zero is $\displaystyle\int_0^z P \, dz$, and this work must be contained in the spring as strain energy (Section 1.7). The strain energy arising due to torsion of the wire is given by equation 2.8 and that due to the bending moment is given by equation 2.18. Thus

$$\int_0^z P \, dz = \frac{T_1^2 l}{\pi G a^4} + \frac{T_2^2 l}{2EAk^2} \tag{2.58}$$

where l is the total length of wire in the spring and is equal to $2n\pi r/\cos \alpha$, since $r/\cos \alpha$ is the radius of a coil, n is the number of coils, and $2a$ is the diameter of the wire. Therefore, substituting for l, for the torques T_1 and T_2, and $\frac{1}{4}\pi a^4$ for Ak^2 for a circular section wire (Figure 2.3),

$$\int_0^z P \, dz = \frac{2nr^3 P^2}{a^4 \cos \alpha} \left(\frac{\cos^2 \alpha}{G} + \frac{2 \sin^2 \alpha}{E} \right) \tag{2.59}$$

Making the assumption than only small extensions are involved, that is, α and r remain constant, then, by differentiating with respect to z which is a function of P,

$$P \, dz = \frac{4nr^3 P}{a^4 \cos \alpha} \left(\frac{\cos^2 \alpha}{G} + \frac{2 \sin^2 \alpha}{E} \right) dP \tag{2.60}$$

Therefore

$$dz = \frac{4nr^3}{a^4 \cos \alpha} \left(\frac{\cos^2 \alpha}{G} + \frac{2 \sin^2 \alpha}{E} \right) dP \tag{2.61}$$

Hence, integrating between $z = 0$ and z, and between $P = 0$ and P,

$$z = \frac{nr^3 P}{a^4 \cos \alpha} \left(\frac{4 \cos^2 \alpha}{G} + \frac{8 \sin^2 \alpha}{E} \right) \qquad (2.62)$$

where z is the extension in the length of the spring under a force P. When the angle 2α between the coils is small, the extension becomes

$$z = \frac{4nr^3 P}{Ga^4}$$

in agreement with equation 2.57 for the extension of a tightly coiled spring.

As the spring is extended, the bending moment on the wire will give the helix a tendency to unwind. Opposing this, the torsion in the wire gives it a tendency to wind up. However, for most metals used in springs, the winding-up tendency will predominate and the spring will coil up as it is extended. It may be shown that a spring with a circular section will coil up or uncoil depending on whether Young's modulus E is greater or less than twice the shear modulus G.

2.6 Stresses in Thin-walled Pressure Vessels

Here we are concerned with vessels containing a liquid or a gas under a pressure p. We wish to know what strains will be produced in the walls of the vessel and how they will depend on the relevant moduli. The vessel shapes that will be considered are either spherical or cylindrical as these are the most commonly used in practice.

Let us consider then the problem of a spherical shell in which the simplifying assumptions will be made that the wall thickness is small (so that the radial stresses are zero and only tangential stress need be considered) and that the weight of the liquid or gas in the vessel may be neglected. The stresses at any point in the shell due to an excess pressure p within the vessel may be resolved into three mutually perpendicular components, one in a radial direction and two in the tangent plane to the shell at that point. These are illustrated in Figure 2.13 for stresses at any point Q. The radial stress, say σ_3, within the shell will be zero if the shell is thin. The tangential stresses, say σ_1 and σ_2, will be equal by symmetry. If now the sphere is imagined to be made up of two halves, the stress holding the two halves together will be σ_1. This stress will act over an area $2\pi rt$, where r is the radius of the sphere and

t the thickness of the shell and where $t \ll r$, so the force pulling the halves together is $2\pi rt\, \sigma_1$. Opposing this is the excess pressure p within the sphere that is trying to force the two halves apart. This will exert a force $\pi r^2 p$ on the hemisphere. Then

$$2\pi rt\, \sigma_1 = \pi r^2 p$$

Therefore

$$\sigma_1 = \frac{rp}{2t} \qquad (2.63)$$

Also, by equations 1.44,

$$\epsilon_1 = \frac{1}{E}[\sigma_1 - \nu(\sigma_2 + \sigma_3)]$$

where ϵ_1 is the strain produced in the direction of the stress σ_1, σ_2 and σ_3 are mutually perpendicular stresses to σ_1, and ν is Poisson's ratio. In

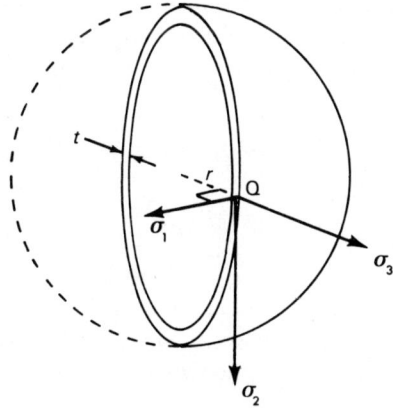

Figure 2.13. Stresses in the wall of a thin-walled spherical pressure vessel

our problem $\sigma_1 = \sigma_2$ and $\sigma_3 = 0$ so, by substituting equation 2.63 in equations 1.44, the tangential strain in the shell is

$$\epsilon = \frac{rp}{2tE}(1 - \nu) \qquad (2.64)$$

A similar treatment may be made for the stresses in a cylindrical shell. As before, the stresses at some point Q in the shell may be resolved

63

into three principal mutually perpendicular stresses σ_1, σ_2, and σ_3, as shown in Figure 2.14: σ_1 is termed the *longitudinal stress*, σ_2 the *hoop stress*, and σ_3 the *radial stress*. If the shell thickness t is small, the radial stress may be neglected in comparison with the other two. The longitudinal stress is determined by imagining the cylinder to be made

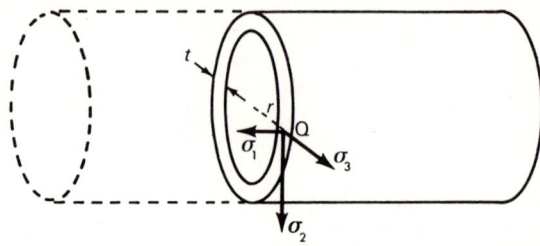

Figure 2.14. Stresses in the wall of a thin-walled cylindrical pressure vessel

up of two parts, as in the figure. The stress σ_1 acts over the annulus of area $2\pi rt$ to hold the two parts together, that is, the force holding them together is $2\pi rt \, \sigma_1$. This is balanced by the excess pressure p within the vessel, which acts against the ends of the tube with a force $\pi r^2 p$. Therefore

$$2\pi rt \, \sigma_1 = \pi r^2 p$$

and

$$\sigma_1 = \frac{rp}{2t} \qquad (2.65)$$

The hoop stress σ_2 may be found by imagining the cylinder to be made up of two longitudinal halves, as shown in Figure 2.15. The stress σ_2 acts over an area $2lt$ to hold the two halves together. Again this is

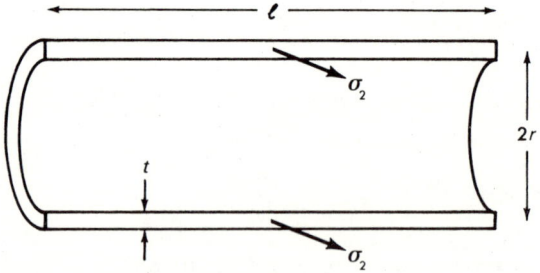

Figure 2.15. Hoop stress σ_2 in a cylindrical pressure vessel

balanced by the excess pressure p which acts on the curved surface with a force $2rlp$, so

$$2lt\,\sigma_2 = 2rlp$$

and

$$\sigma_2 = \frac{rp}{t} \qquad (2.66)$$

The hoop stress σ_2 is thus twice the longitudinal stress σ_1.

If ϵ_1 and ϵ_2 are the longitudinal and hoop strains respectively, corresponding to the stresses σ_1 and σ_2, then by equations 1.44

$$\epsilon_1 = \frac{1}{E}[\sigma_1 - \nu\,(\sigma_2 + \sigma_3)]$$

$$\epsilon_2 = \frac{1}{E}[\sigma_2 - \nu\,(\sigma_1 + \sigma_3)]$$

and therefore, since $\sigma_3 = 0$, the longitudinal strain is

$$\epsilon_1 = \frac{rp}{2tE}(1 - 2\nu) \qquad (2.67)$$

and the hoop strain

$$\epsilon_2 = \frac{rp}{2tE}(2 - \nu) \qquad (2.68)$$

2.7 Vibrations of Stressed Bodies

If a stressed body such as a simple cantilever carrying a load at its free end is temporarily displaced from its equilibrium condition, restoring forces will come into play causing the body to vibrate. For bodies of simple shape and under simple stress conditions the period of vibration can be readily determined, but for the general problem the treatment may become very difficult.

For the cantilever the period of vibration may be determined as follows. Let us consider the simple cantilever of length l supporting a mass m as shown in Figure 2.16. The deflection of the end of the rod, where $x = l$, caused by the load mg is

$$y = \frac{mgl^3}{3EAk^2}$$

65

by equation 2.24. In this equilibrium condition there is a restoring force F acting, where

$$F = mg = \frac{3EAk^2}{l^3} y \qquad (2.69)$$

If now the end is depressed by a small amount, the restoring force will act on the mass m to produce a restoring acceleration $-\mathrm{d}^2y/\mathrm{d}x^2$.

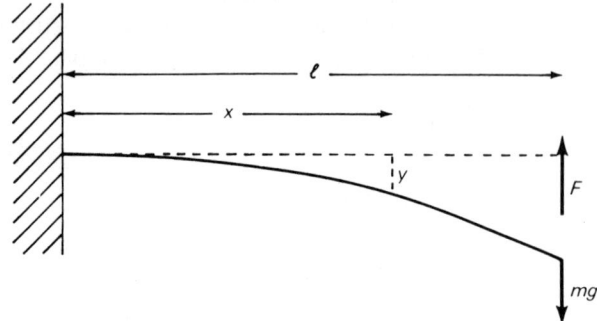

Figure 2.16. Simple cantilever supporting a mass

That is, the restoring force F is given by

$$F = m\frac{\mathrm{d}^2y}{\mathrm{d}x^2} = -\frac{3EAk^2}{l^3} y \qquad (2.70)$$

But

$$\frac{\mathrm{d}^2y}{\mathrm{d}x^2} + \frac{3EAk^2}{ml^3} y = 0 \qquad (2.71)$$

is the equation of a simple harmonic motion whose period of vibration is

$$T = 2\pi \sqrt{\frac{ml^3}{3EAk^2}} \qquad (2.72)$$

This is, of course, only an approximation of the practical case since, for example, the effects of the mass of the cantilever and of air resistance have been neglected.

We can also find the period of axial vibration of a helical spring. To do this we must derive expressions for the kinetic energies of the vibrating spring and its suspended load as well as for the potential energy of the system and see how the total energy varies with time. If the suspended load exerts a force $P = Mg$ on the spring (Figure 2.17) and if the vibrating end is moving with a velocity dz/dt when the extension is z, then the kinetic energy of the load is $\frac{1}{2}M(dz/dt)^2$.

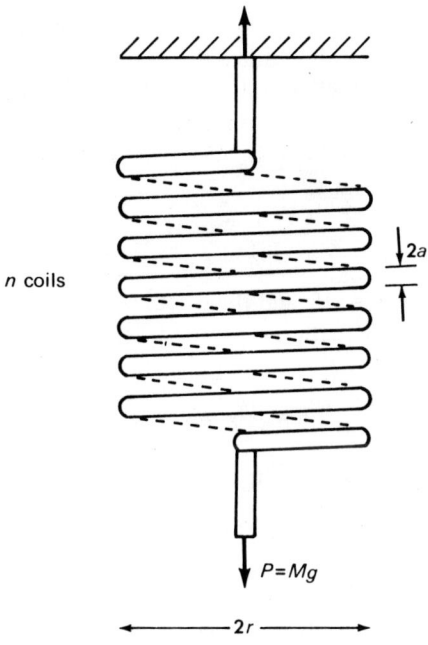

n coils

2a

P=Mg

2r

Figure 2.17. Axial vibration of a helical spring

Similarly, if an element of length δs and distance s from the fixed end of the spring is considered, this element will have a kinetic energy

$$\frac{1}{2}m_0 \, \delta s \left(\frac{s}{l}\frac{dz}{dt}\right)^2$$

where m_0 is the mass per unit length of the spring and $l = 2\pi nr$ is the length of the spring. The kinetic energy of the whole spring is therefore

$$\frac{m_0}{2l^2}\left(\frac{dz}{dt}\right)^2\int_0^l s^2 \, ds = \frac{1}{6}m_0 l\left(\frac{dz}{dt}\right)^2$$

$$= \frac{1}{6}m\left(\frac{dz}{dt}\right)^2 \tag{2.73}$$

where m is the mass of the spring. The kinetic energy of the combined spring and load is therefore

$$\frac{1}{2}M\left(\frac{dz}{dt}\right)^2 + \frac{1}{6}m\left(\frac{dz}{dt}\right)^2 = \frac{3M+m}{6}\left(\frac{dz}{dt}\right)^2 \qquad (2.74)$$

To determine the potential energy of the system we can make use of the work of Sections 2.2 and 2.5 since the strain energy will be equal to the potential energy. If we consider only the approximate solution to the spring problem, that is, assuming a closely wound helix so the plane of a coil is normal to the spring axis (that is, $\alpha = 0$), then the strain energy due to torsion alone is

$$W = \frac{T^2 l}{\pi G a^4} = \frac{2nr^3 M^2 g^2}{G a^4} \qquad (2.75)$$

where the torque $T = rP = rMg$ and the length of the spring $l = 2\pi nr$, as in Section 2.5. Also, by equation 2.57, the extension under a load $P = Mg$ is

$$z = \frac{4nr^3}{G a^4} Mg \qquad (2.76)$$

Hence the potential energy, equal to the strain energy, is

$$W = \frac{G a^4}{8nr^3} z^2 \qquad (2.77)$$

Now the total energy of the vibrating spring and its suspended load, the sum of the kinetic and potential energies, must remain constant, so

$$\frac{3M+m}{6}\left(\frac{dz}{dt}\right)^2 + \frac{G a^4}{8nr^3} z^2 = \text{const.} \qquad (2.78)$$

If we now differentiate this with respect to time t,

$$\frac{3M+m}{3}\frac{d^2z}{dt^2} + \frac{G a^4}{4nr^3} z = 0 \qquad (2.79)$$

This is an equation for simple harmonic motion of period T, leading to the result that the period of vibration of a circular-section wire helical spring of n turns of radius r and mass m, supporting a mass M, is

$$T = 2\pi \sqrt{\frac{(3M+m)/3}{G a^4 / 4nr^3}}$$

$$= 4\pi \sqrt{\frac{nr^3 (3M+m)}{3G a^4}} \qquad (2.80)$$

where $2a$ is the diameter of the wire.

2.8 Measurement of the Moduli

It is not normally possible to measure stress directly, and therefore it is usual to apply known stresses and to measure the resulting strain. This is fairly easy provided the body under test is of simple geometric shape so that the direction and magnitude of the measured strain is clearly defined in relation to the applied stress. Using direct mechanical measurement, the strain can generally be measured to an accuracy no better than about one in 10^4 or 10^5.

Young's modulus is readily obtained from a stress–strain curve drawn for the body whilst it is subjected to tensile stresses. For a wire, Searle's method is the best. In its usual form the apparatus consists of two rectangular frameworks constrained to move parallel by hinged cross-links, as shown diagrammatically in Figure 2.18. The frameworks

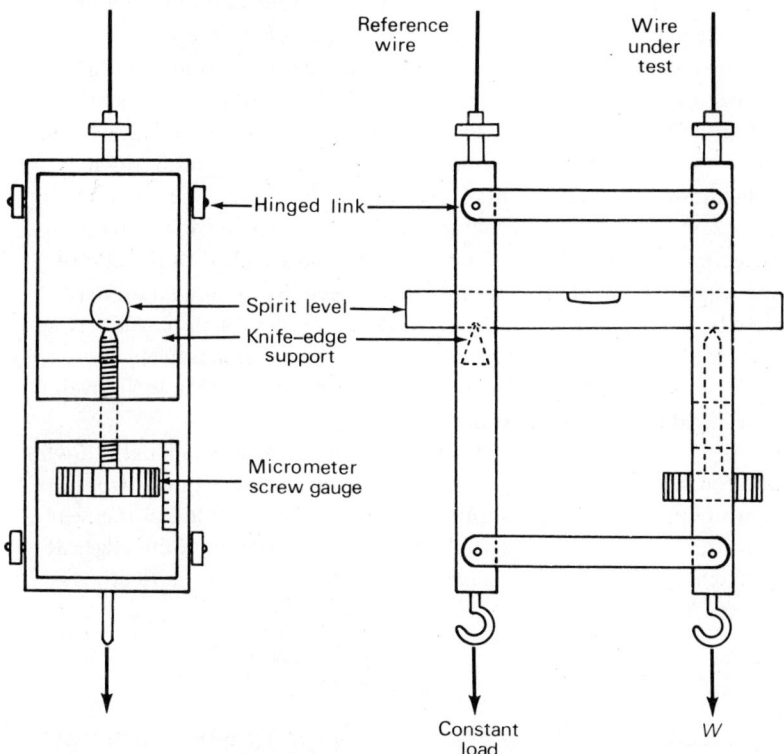

Figure 2.18. Searle's apparatus, shown diagrammatically, for the determination of Young's modulus for a wire

are supported by two hanging wires, one of which is the wire under test which can be stretched by means of various loads W and the other is a wire of the same material. This wire acts as the reference length and is kept taut by a hanging constant load. Between the two frameworks is a spirit level which may be adjusted to be horizontal by a micrometer gauge and screw as shown in the figure. Increased loads on the test wire cause it to extend, lowering the right-hand framework and the end of the spirit level. The spirit level is adjusted by the micrometer screw to be horizontal again, and the increase in length is determined by the change in the micrometer gauge reading. The extensions may be determined for various loads, and, from measurement of the original length and mean cross-sectional area, a stress–strain curve may be drawn; from the slope of the linear part (that is, over the elastic region), Young's modulus is obtained. A reference wire of the same material is used so that changes in length due to temperature variations of the surroundings will be cancelled out, as will the effect of any give in the top support.

For rod specimens the stress may be applied by various mechanical means, for example by heavy weight and lever. If the weight is moved along the lever by a motor, the test conditions may be cycled mechanically, so increasing and decreasing the stress at a rate and frequency which may be varied at will. The change in length may be measured by a micrometer dial gauge or alternatively by the change in electrical capacity between a fixed metal plate and one attached to the end of the test rod when their separation is changed. Stress and strain may then be plotted automatically by a chart recorder. Such tensile-testing machines are commercially available and, since comparatively short specimen rods are used, it is easy to enclose the specimen in an oven to vary the temperature conditions of the test.

As well as from the direct measurement of changes in length of wire and rod specimens, Young's modulus can also be determined from measurements of the deflection of beams. From Section 2.3 it is seen that, if a light beam of length $2l$ is freely supported by knife edges at each end and deflected by a central load of mass m, the deflection at the centre will be

$$y = \frac{mgl^3}{6EAk^2}$$

by equation 2.32, where $Ak^2 = bd^3/12$ for a rectangular section beam of breadth b and depth d (Figure 2.3). The deflection may be determined by a variety of methods, using, for example, a microscope, an

70

optical lever, a micrometer dial gauge, the capacitance between plates, or optical interference methods. A beam method is especially useful for measuring Young's modulus for crystalline 'whiskers'. These are too small for direct extension measurements but can be mounted as a cantilever beam and a small load applied to the free end. The depression of the end may be measured with a microscope with a calibrated graticule or vernier eyepiece, and Young's modulus calculated from equation 2.24:

$$y = \frac{mgl^3}{3EAk^2}$$

where l is the free length of the whisker and $Ak^2 = \pi a^4/4$ where $2a$ is its diameter.

Other indirect methods have been used for the determination of Young's modulus. For example, in the previous section we derived an expression for the period of vibration of a simple cantilever (equation 2.72) from which it could be determined, but on the whole such methods are not particularly accurate as either inaccurate measurements are involved or the theory is not sufficiently precise. Therefore these indirect methods should be used only if the specimen is of some special shape, such as a coiled spring, or if it is not readily available for direct experimentation as, for example, a specimen inside a vacuum chamber. Here it may be easier to make the specimen vibrate and to measure its period than to measure a strain or displacement.

The shear modulus is not so easy to measure directly. If a cube of the material is sheared according to the definition of the modulus, it is not possible to measure the angle of shear with any precision. If, however, the specimen is in the form of a wire or rod, then it may be twisted and the shear modulus G determined from equation 2.5, that is,

$$T = \frac{\pi G r^4}{2l} \theta$$

by measuring the couple T required to twist one end of the specimen of length l and radius r through an angle θ. For a wire the couple can conveniently be applied as shown in Figure 2.19. The torque is applied by the cords acting tangentially to a metal cylinder attached to the end of the wire. This also keeps the wire taut. The twist over the test length l is measured using each of the two mirrors attached to the wire in conjunction with the usual lamp and scale arrangement.

For rod specimens the same principle is used as for a wire but with a more robust arrangement. As for tensile testing, commercial machines are available for applying torsion to a rod and measuring the angle of twist, both readings being taken automatically and fed into a chart recorder. Again the tests may be cycled by varying the rate, magnitude, and frequency of the torsion, as well as by varying the temperature. Other indirect methods have been used to determine the shear modulus as, for example, those involving the extension of a helical spring (equations 2.57 and 2.62) or its period of vibration (equation 2.80).

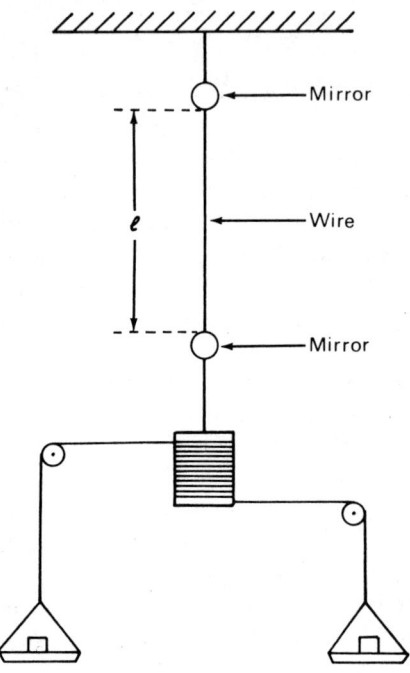

Figure 2.19. Illustrating the method for determining the shear modulus of a wire specimen

The measurement of the bulk modulus also presents problems. It may be calculated from equation 1.55

$$E = \frac{9KG}{3K + G}$$

from a knowledge of Young's modulus E and the shear modulus G. This however has disadvantages. A wire specimen, owing to its past heat treatment and drawing, will probably have very different properties from a bulk specimen or from a rod that has been machined. Thus the specimens used in the measurement of Young's modulus and the shear modulus may lead to a considerable discrepancy between the calculated value of the bulk modulus and its true value.

There are no direct methods of measuring the bulk modulus since all involve the pressure being applied hydrostatically with consequent corrections for the compressibility of the fluid or the change in volume of the containing vessel. Indirect methods of measurement have been used by Bridgman. His apparatus is shown diagrammatically in Figure 2.20. The specimen is in the form of a rod which fits into a steel

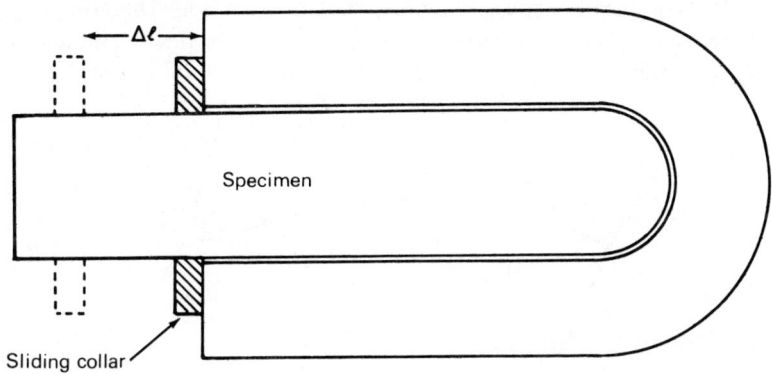

Figure 2.20. *Apparatus, shown diagrammatically, for the determination of the bulk modulus of a solid specimen*

cylinder, and the whole is then subjected to a hydrostatic pressure. The specimen rod contracts relative to the steel block, so the sliding collar moves along the rod. When the pressure is released, the collar remains at its new position on the rod and hence the movement of the collar Δl measures the change in length of the rod relative to the steel cylinder. The true contraction is then known providing the change in length of the cylinder is known. This method actually measures the longitudinal strain; the volume strain is three times this. Now, the pressure being known, the bulk modulus may be calculated. As well as sliding collars, sliding contacts on electrical resistance wires have been used to measure the changes in length. The method, however, is dependent on a knowledge of the absolute compressibility of the cylinder.

The other related parameter to be determined is Poisson's ratio. This can be measured directly by methods similar to those for Young's modulus. For a particular longitudinal strain the corresponding lateral strain is found by measuring the diameter of the wire or rod with a micrometer screw gauge. Poisson's ratio can also be determined from measurements on a bent beam of rectangular cross-section. We have already shown that the longitudinal strain at a distance δr from the neutral surface is $\epsilon = \delta r / r_1$, where r_1 is the radius of curvature of the neutral axis (equation 2.10). Also the beam material on the convex side will be stretched. With this longitudinal extension is associated a lateral contraction. Conversely, on the concave side there is a longitudinal compression and a consequent lateral expansion. Therefore there is a bending of the beam in the plane normal to the longitudinal plane, as shown in Figure 2.21. This anticlastic curvature will have a

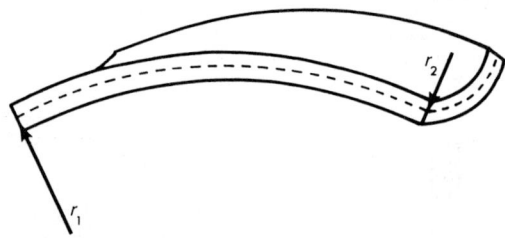

Figure 2.21. Anticlastic curvature of a beam

radius r_2 and the lateral strain of the beam at a distance δr from the neutral surface will be $-\delta r / r_2$. Poisson's ratio then will be

$$\nu = -\frac{\text{lateral strain}}{\text{longitudinal strain}}$$

$$= \frac{\delta r / r_2}{\delta r / r_1}$$

$$= \frac{r_1}{r_2} \qquad (2.81)$$

The radius r_1 may be determined from measurements of the sag of the curve; r_2, the anticlastic curvature, may be found by means of mirrors fixed to the sides of the beam to form optical levers in conjunction with lamps and scales, or by clamping pointers to the beam and making measurements of the distances moved by their ends to determine the angles traversed, or by optical interference methods. The last involve laying a flat glass plate on the beam, which must be of highly polished metal, plastic, or glass, to form hyperbolic interference fringes, or even resting a lens on the surface to form Newton's rings. Elliptical fringes will then be formed. If the beam is bent so that a radius of curvature is equal to that of the lens, straight line fringes will be formed. From measurements of the fringe shape and separation, the required radii may be determined and Poisson's ratio found from equation 2.81.

In practical structures it is often necessary to know the stresses occurring at different points to check assumptions made in the theoretical design studies or to see if critical stresses are being reached. As was said earlier, it is not usually possible to determine these stresses directly and instead it is necessary to determine the strain and calculate the stress from the appropriate modulus. As well as the methods of strain measurement already discussed, other methods are used. The most important of these uses strain gauges, a discussion of which will be deferred to the next section.

A method involving photoelastic analysis may be used in the design of complicated mechanical structures which are to be subjected to strain. In this method a model of the part to be studied is made from transparent plastic and placed between crossed optical polarisers. On looking through the assembly, only a dark field is seen. The plastic model is now subjected to forces similar to those which would be applied to the actual part under use. Light- and dark-coloured areas are seen in the model as molecules of the plastic are distorted by the applied stress and the plastic becomes birefringent, acting as a retardation plate. The amount of birefringence introduced at each point in the model is a measure of the strain occurring at that point, and the two axes of the retarder at each point are parallel and perpendicular to the directions of the strain at that point. Thus the points of greatest strain are points of greatest retardation, resulting in the greatest mismatch of polarisation direction with that of the analyser and in most light being transmitted. The measurement of the actual strain in a particular region may be made in several ways. Calibrated retarders may be used to cancel out the retardation introduced by strain in the model and hence give a measure of the retardation introduced. From subsidiary experiments on models of simpler shape the relation between strain and retardation, that is, the stress-optic constant, may be determined. Alternatively, it is possible to determine the retardation from the colour of the fringe at a particular point, since retardation depends on wavelength and the assembly is illuminated by white light. This method, however, is at its most useful when it is used to locate the points of maximum strain, rather than for determining actual values, so that the structure may be strengthened at those appropriate points.

Another method involves the coating of the body under test with a brittle lacquer. This method has the disadvantage of being relatively insensitive as well as being able to indicate only tensile strains, although it does have the advantage that it can indicate the direction of the maximum tensile stress.

X-ray methods have an advantage in that they can determine the elastic component only of the strain, but the sensitivity is not as high as the direct mechanical methods and much bulky and expensive equipment must be used, although this may not be an especial disadvantage as such equipment will normally be found in a modern research laboratory.

2.9 Strain Gauges

Strain gauges make use of the property of certain materials that the change in their electrical resistance is proportional to the strain applied. The most common materials used are metal alloys, which will thus form the subject of this section.

If wires of these alloys are fixed to the surface of a body under test, they will be strained by the same amount as the body at that point, and, from measurements of resistance change, the magnitude of the strain may be determined. Alternatively, the strain gauges may be built into a structure, for example they may be embedded in the concrete foundations of a building, the structure of a bridge, or under road surfaces, so that changes in strain with time or condition of service may be monitored. They have the advantage that it is the strain at a very localised point that is determined and, by using suitably shaped gauges, the direction of the strain may also be determined. As well as linear strain they can be used to measure torque, pressure, acceleration, and vibration.

The gauges are in the form of a grid of fine wire or, more commonly, a grid of foil mounted on an insulating membrane for sticking to a surface or embedded in an insulating medium for casting into a concrete structure. Common configurations are shown in Figure 2.22. The thickness of the metal foil, which is usually of cupro-nickel or nickel–chromium alloy, varies from about 0·01 mm (0·0004 in) to 0·025 mm

(0·001 in). The medium on which it is mounted or embedded sets the upper limit of temperature at which the gauges may be used, but they are generally commercially available for use up to about 800°C. Various adhesives are available for sticking the gauge to the body under test, for example Araldite is commonly used. This is an epoxy resin that may

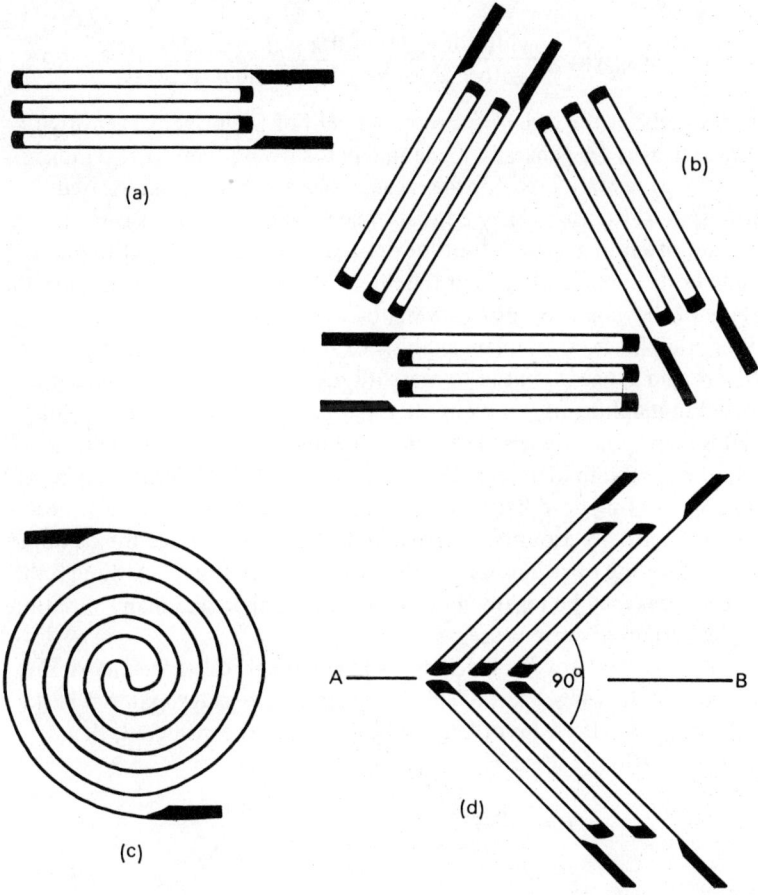

Figure 2.22. Strain gauges: (a) linear, (b) rosette, (c) diaphragm, and (d) torque

be cured by the heat from a lamp held near. Commercially available gauges have resistances of about $50-1000\,\Omega$ depending on the size and type.

The linear gauge in Figure 2.22 is used for measuring strain along the direction of the gauge, and for this reason the grid reversing points are thickened to reduce the effect of cross-sensitivity, that is, the sensitivity to strain across the grid at right angles to the active direction or direction of principal sensitivity. The cross-sensitivity may be defined as

$$\text{cross-sensitivity} = \frac{\text{width of gauge grid} \times \text{principal sensitivity}}{\text{active gauge length} \times \text{number of wires}} \quad (2.82)$$

It is usually of the order of several per cent of the principal sensitivity. Linear gauges are generally made in lengths from 5 mm to 150 mm. Rosette gauges are used for the measurement of more complicated strains in a structure. They consist of some arrangement of three linear gauges, often in a 'crow's foot' or delta format, and are used to determine both the magnitude and the direction of the principal strain in the plane of the gauge. By using several of these mounted in different direction planes, a complete analysis of the strain may be made. Diaphragm gauges are used to measure the strains and deflections produced in diaphragms, for example they may be mounted on the end-plates of pressure vessels. They are generally available from about 10 mm to 70 mm diameter. Torque gauges are for the measurement of torsion and for shear. They are mounted so that AB (see figure) runs circumferentially around the shaft, and with correct usage they eliminate the reading of strains due to the bending of the shaft. Although the above types may be regarded as the main ones there are many variations in the designs in current use.

The relation between the changes in electrical resistance and strain is given by the gauge factor. This is related to the gauge material in the following way. If a wire of length l and radius r is considered, then Poisson's ratio for this is

$$\nu = -\frac{\text{decrease in width/unit width}}{\text{increase in length/unit length}}$$

$$= -\frac{\delta r/r}{\delta l/l}$$

$$= -\frac{\frac{1}{2}\delta A/A}{\delta l/l} \quad (2.83)$$

where A is the cross-sectional area of the wire. This last step follows simply since $\delta A/A = 2\pi r \delta r/\pi r^2 = 2 \delta r/r$ (Figure 2.23). Also the resistance of this length of wire will be

$$R = \frac{\rho l}{A} \qquad (2.84)$$

where ρ is the resistivity of the material. Therefore, taking logarithms and differentiating,

$$\frac{\delta R}{R} = \frac{\delta \rho}{\rho} + \frac{\delta l}{l} - \frac{\delta A}{A}$$

$$= \frac{\delta \rho}{\rho} + \frac{\delta l}{l} (1 + 2\nu) \qquad (2.85)$$

from equation 2.83. If an assumption is made that the resistivity ρ remains constant, that is, $\delta \rho / \rho = 0$, so that the change in resistance is

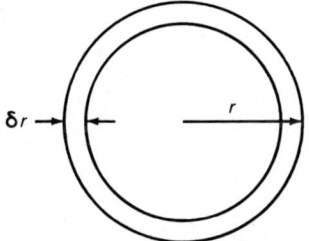

Figure 2.23.   Cross-section of a
wire of radius r

purely a function of the dimensional changes, then the gauge factor or strain sensitivity which by definition is

$$\text{gauge factor} = \frac{\delta R/R}{\delta l/l} \qquad (2.86)$$

$$= 1 + 2\nu \qquad (2.87)$$

from equation 2.85. Since ν is about 0·3 for the materials used in alloy strain gauges, equation 2.87 implies that the gauge factor should be about 1·6. In practice, however, the gauge factor is usually in the range of about 1·9—2·3 and therefore the assumption that the resistivity is independent of the applied stress is not strictly true. It is thus necessary to measure the gauge factor. The magnitude of the strain will then be

$$\epsilon = \frac{\delta R}{R} \times \frac{1}{\text{gauge factor}} \qquad (2.88)$$

Since the gauge can only be used for measuring strains once it has been fixed to the body or structure under test, and once fixed down it

cannot normally be removed without damage, it is necessary to determine the gauge factor for one or two gauges and assume that all the
other gauges manufactured in that batch will behave in the same way.
The calibration of the gauge is best done by fixing it to the surface of a
beam and flexing it. Preferably two gauges should be used, one on each
side of the beam so that one is contracted as the other is expanded. If

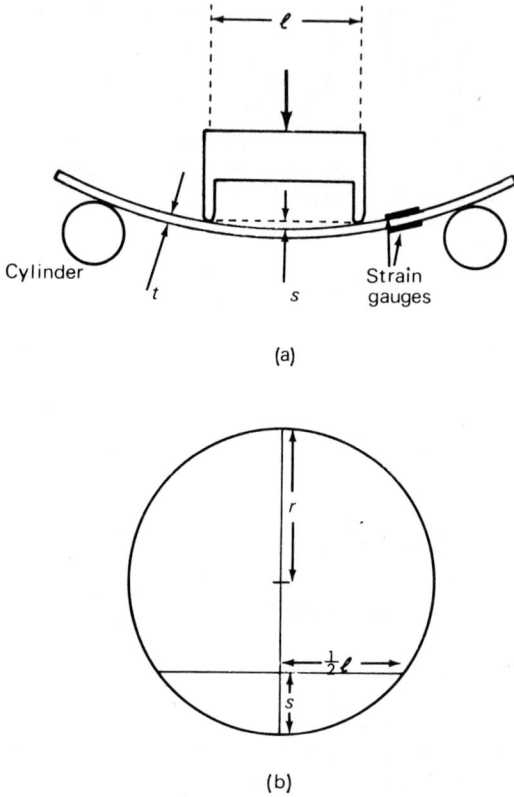

(a)

(b)

Figure 2.24. Calibration of strain gauges by the flexing beam method

these are connected to adjacent arms of a Wheatstone bridge, the
sensitivity of measurement will be doubled. The beam is bent into an
arc as shown in Figure 2.24, and the sag s over the distance l is measured
by means of a dial gauge. It is important that there should be no angular
twist along the beam as it is deformed. The radius of curvature r is

calculated using the rectangular property of chords of a circle theorem
[Figure 2.24(b)]:

$$(2r - s)\, s = \left(\tfrac{1}{2}l\right)^2$$

and therefore

$$r = \frac{l^2}{8s} \qquad (2.89)$$

neglecting the product of small quantities. Also, from equation 2.10,
the linear strain of a longitudinal element distance δr from the neutral
plane of a bent beam of curvature r is

$$\epsilon = \frac{\delta r}{r}$$

Hence, substituting from equation 2.89 where the thickness of the beam
is $t = 2\,\delta r$, the linear strain is

$$\epsilon = \frac{\delta l}{l} = \frac{4st}{l^2} \qquad (2.90)$$

If it is suspected that the gauge may behave differently when under
tension from when under compression, only one gauge should be used
and the beam turned over for the other condition.

Having determined the strain and corresponding change in resistance
$\delta R/R$, the gauge factor is calculated from equation 2.86 and this value
applied to the other gauges in the same batch. It is found that variations
of the gauge factor within a batch depend on the gauge size and resistance
and remain constant for up to about 0·2% strain. For small low-resistance
gauges the variations are usually less than ± 2·5%, dropping to better
than ± 0·5% variation for the long high-resistance gauges.

Strain gauges are usually used for the measurement of fairly small
strains since in practice strains occurring in safe structures are generally
small, perhaps only several parts in 10^6, and consequently the changes
in resistance are only of this order. It is therefore necessary to use fairly
sensitive bridge arrangements for the measurements of resistance.
Conversely, strain gauges are liable to mechanical failure for strains
greater than about 0·5%. Use of the gauges may be divided into two
main categories, static or dynamic.

When the gauge is used statically, measurement may be by a direct
deflection method or a null method. In the direct deflection method a
calibrated galvanometer is used in a bridge circuit, such as in Figure 2.25.

The measuring gauge G is in one arm and an exactly similar gauge C used as a temperature compensator is in the adjacent arm. The other arms should have similar-valued resistances. Variations in resistance due to temperature variation are important since the gauge material will probably have a temperature coefficient of about 15 parts in 10^6 per

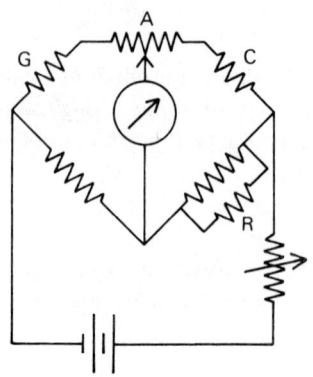

Figure 2.25.   Strain gauge bridge suitable for a direct reading method of use

degree Celsius at room temperature, but this variation is almost completely cancelled out by using a compensating gauge. This should be under the same ambient temperature conditions as the measuring gauge G but is unstrained. In many applications it is possible to double the sensitivity by mounting this compensating gauge in a position where it will be strained by an equal but opposite amount of strain to the main gauge whilst still under the same ambient temperature conditions, such as when mounted on opposite sides of a beam. Zero balancing of the bridge may be done by using an apex resistor A, and calibration is achieved by shunting one of the bridge arms with a high resistance R. The current through the strain gauges should not exceed about 10 mA for long-term use or about 100 mA for short periods to avoid self-heating effects which will alter the resistance. These values of current depend very much of course on the particular gauge and its use. For long-term testing it may be necessary to make corrections for zero drift due to creep or even due to the fixing cement absorbing moisture and swelling.

For a null method measurement of the strain gauge resistance, the circuit shown in Figure 2.26 is suitable. This uses a temperature-compensating gauge C as well as the measuring gauge G and an apex resistor for zero balancing as before. Readings of percentage change in

82

resistance are made from the slide wire of known resistance per unit length, which is fitted with shunts to give different ranges.

Dynamic tests are usually made over short periods and it may not be necessary to make compensation for temperature variation. If a current is passed through the gauge and the change in voltage across it due to

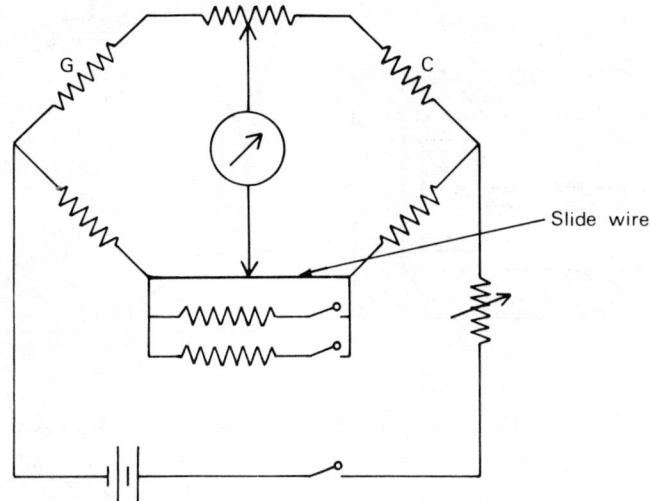

Figure 2.26. Strain gauge bridge suitable for null method measurements

the strain pulse displayed on an oscilloscope, the magnitude and length of the pulse may be measured from the trace, preferably after photographing. The measurement of alternating strains of frequencies as high as 100 kHz is possible using strain gauges.

As well as the direct measurement of strain, strain gauges have many other applications. For example, if they are mounted on a cantilever carrying a heavy end-load, the system may be used as an accelerometer (Figure 2.27) since any acceleration will cause the beam to bend owing to inertia and affect the gauge resistance.

By fitting gauges to a proving ring, a system for weighing heavy loads may be made. Such a system may be used in a weighbridge or in a large tensile testing machine, for example. A proving ring is simply a ring of steel, as shown in Figure 2.27(b); an applied load will cause it to distort into an elliptical form. The system may be calibrated so that the change in resistance of the gauges is related to the applied load.

Photoelastic strain gauges are also used. The principle is the same as that for the photoelastic stress-analysis method already mentioned but, instead of making a model in plastic of the body under test, the plastic is coated onto the body as shown in Figure 2.28. The clear plastic is

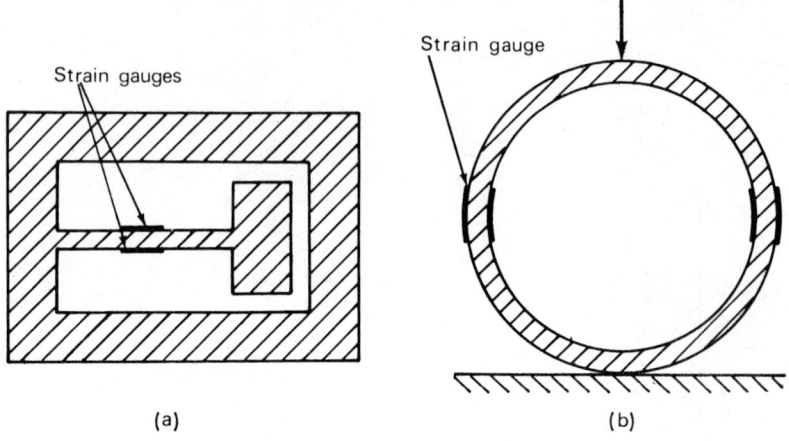

Figure 2.27. Use of strain gauges to measure (a) acceleration, and
(b) heavy loads

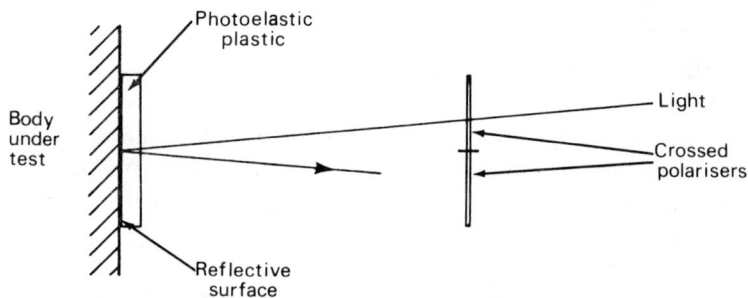

Figure 2.28. Use of a photoelastic strain gauge

cemented into position with a reflecting cement. Since the light passes twice through the plastic, the sensitivity for a given thickness is doubled. As the body is strained, the plastic film also becomes strained and birefringent, giving a relative retardation between the two rays produced within the plastic. The amount of birefringence and consequent relative retardation is a measure of the strain, as already discussed in the previous section. Photoelastic strain gauges have an advantage over resistance

84

strain gauges in that they examine the strain over an area rather than at a point. Since the plastic is equally sensitive to strain in all directions in its plane, each gauge gives information similar to that obtained by using an infinite number of zero-length gauges. A disadvantage is that it can measure only surface strains that are also accessible to light. These gauges are also less sensitive and more difficult to calibrate than resistance strain gauges. Consequently, their greatest use is in determining the position of any very large stress concentration so that, by a modification in design, it can be reduced to a more tolerable level.

2.10 Liquids

So far we have discussed only the elasticity of solids, but liquids also have some similar properties. Of interest is the tensile strength of a liquid column since it concerns, for example, the rise of sap in trees and the formation of bubbles in supersaturated liquids. How much stress or strain can a liquid column stand before it separates? The main work has been on water, but unfortunately there is considerable discordance between results. This may be due to a variety of causes, such as the amount of dissolved gas, and there is even a possibility that rupture may occur first at the container wall rather than in the bulk of the liquid, so the adhesive properties between the liquid and the container wall will have an effect.

One method of measuring the breaking strain of a liquid is to contain it in an evacuated thick-walled glass tube sealed at the ends. The liquid is then heated so that it expands to fill the tube. On cooling slowly the liquid column will try to contract and will finally break to form a bubble of vapour. From the length of the tube and length of this bubble the maximum volume strain may be determined.

Stress has been applied by rapidly rotating a tube containing liquid. If the tube is U-shaped and contains liquid in one arm and liquid in equilibrium with its own vapour in the other, as shown in Figure 2.29, and is rotated in its own plane about a point A, a tension will exist in the liquid owing to centrifugal forces. The tension will be a minimum at B and a maximum at C. Hence, by increasing the speed of rotation, the breaking stress may be determined from a knowledge of the centrifugal force applying at the breaking point.

Studies of the compressibilities or bulk moduli of liquids have been carried out by Bridgman and other workers. Bridgman used apparatus similar to that which he used for the determination of the bulk moduli of solids. For example, the apparatus of Figure 2.20 can be modified to contain a liquid which is retained by a piston. It has been found that in the range from −30°C to 200°C mercury is the least compressible of

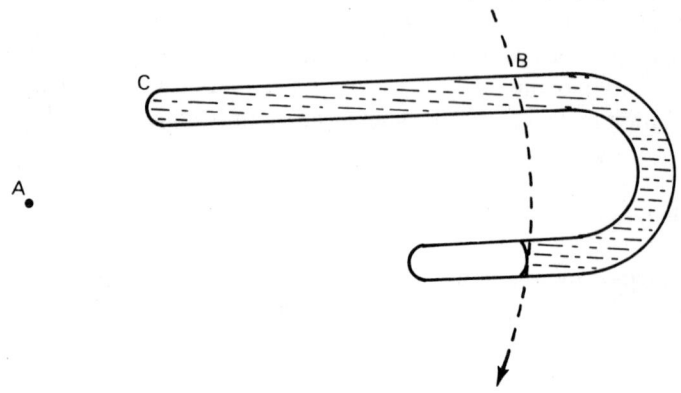

Figure 2.29. U-shaped tube for determining breaking stress of a liquid

all the liquids with glycerine being the least compressible non-metal liquid. For pressures above $10^9 \, \text{N m}^{-2}$ (10 000 atm) the relative volume change becomes about the same for all liquids.

Measurements of the tensile properties of water are of great interest in work on cavitation. Cavitation is the formation of voids or bubbles of liquid vapour within the liquid owing to localised rupturing of the liquid. This has the effect, for example, of reducing the efficiency of ships' propellers, producing drag and instability of underwater missiles, and causing divers' bends.

The inception of cavitation appears to be with the growth of undissolved vapour or gas which may be free in the liquid or attached to foreign particles. A cavity may grow in a liquid and attach itself or remain close to the moving body causing the cavitation and may reach an equilibrium size. Two distinct types of cavitation occur. In the first, termed bubble or transient cavitation, small bubbles suddenly appear on the solid boundary past which the liquid is flowing. These grow to a certain size and then collapse again. In the other type, termed steady state or sheet cavitation, the cavities form on the solid boundary face and remain attached. They will stay in this steady state condition until the flow rate of the liquid over the boundary is changed.

As well as causing loss of efficiency, for example in the loss of thrust in ships' propellers and hydraulic turbines, cavitation can cause erosion. The bubbles may grow to their full size in about 2×10^{-3} s, which is too short a time for much air or other dissolved gas to come out of solution, so these bubbles are in fact highly evacuated. If they subsequently collapse in the same order of time, the liquid rushes in with great velocity. If the bubbles form on a solid boundary surface, very great localised stresses may be caused by the impinging liquid as the bubbles collapse. This can cause localised fatigue failure and consequent erosion after only short periods.

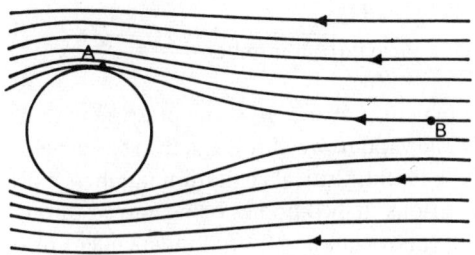

Figure 2.30. Flow of a liquid past a solid body

If we consider a liquid flowing past a sphere, as shown in Figure 2.30, where A and B are points on the same streamline, then according to Bernoulli's theorem which will be derived in Chapter 4

$$\frac{1}{2} \rho v_A^2 + p_A = \frac{1}{2} \rho v_B^2 + p_B \qquad (2.91)$$

where ρ is the density of the liquid, assumed constant along a streamline, and v_A, v_B, p_A, and p_B are the velocities and absolute pressures of the liquid at A and B respectively. Then

$$p_A = p_B - \frac{\rho v_B^2}{2}\left[\left(\frac{v_A}{v_B}\right)^2 - 1\right] \qquad (2.92)$$

Now, from experiment it is found that the condition for the onset of cavitation is that

$$p \leqslant p' \qquad (2.93)$$

where p is the minimum pressure at any point on the solid–liquid boundary and p' is the vapour pressure of the liquid at the prevailing

87

temperature. Conversely, the condition for the avoidance of cavitation is that

$$p > p' \qquad (2.94)$$

If p is the pressure at the solid–liquid interface A in Figure 2.30, that is, $p = p_A$, then, combining equation 2.94 with equation 2.92, the condition for the avoidance of cavitation is that

$$\frac{2(p_B - p')}{\rho \, v_B^2} > \left[\left(\frac{v_A}{v_B} \right)^2 - 1 \right] \qquad (2.95)$$

The left-hand term is the *cavitation number*, that is,

$$\text{cavitation number} = \frac{2(p - p')}{\rho v^2} \qquad (2.96)$$

where p is the ambient pressure, p' is the pressure within a cavity (approximately the vapour pressure), v is the stream velocity, and ρ is the density of the liquid. Critical cavitation numbers exist for the cavitation caused by various shaped bodies being moved through a liquid. The right-hand term of equation 2.95 is independent of the stream velocity since $v_A \propto v_B$ for a particular shaped body. For simple-shaped bodies the right-hand term may be calculated, but generally it must be determined experimentally.

Chapter Three

PLASTICITY

3.1 Introduction

In Chapter 1 we discussed the shape of stress–strain curves for some typical materials. These were mainly metals as it is in these that the main interest in elastic properties lies. In this chapter we shall again concentrate mainly on the properties of metals. The stress–strain curves discussed generally exhibited some elastic region in which the strain was proportional to the applied stress, so when the stress was reduced to zero then so also was the strain, at least up to a certain value of the stress. Beyond this yield point, plastic deformation occurred and the test specimen did not completely recover its original dimensions. It is the reasons for this permanent deformation and their implications that constitute the subject matter of this chapter. In the same way as we had to resort to a discussion of the forces between individual atoms to give some explanation of elasticity, we must again resort to a discussion at the atomic level to account for the plastic properties of a material. To do this it is necessary first to recall our knowledge of the crystallographic structure of materials, and especially metals. We can then go on to discuss the defects that may occur in those structures and to what extent these defects may affect the bulk properties of the material.

3.2 Structure of Solids

As a metal, whether it be a pure metal or a mixture of metals forming an alloy, freezes out from its liquid state, its atoms undergo a long-range ordering to form a definite crystallographic structure. In the common engineering metals the ordering process is such as to cause the atoms to pack closely together in three possible ways, depending on the particular metal atoms involved. These are the face-centred cubic (f.c.c.), body-centred cubic (b.c.c.), and close-packed hexagonal (c.p.h.) forms.

89

Figure 3.1 shows the unit cells for these crystallographic types. A unit cell may be defined as the smallest repeating volume which has a lattice point at each of its corners. Thus a complete crystal comprises a three-dimensional array of these unit cells.

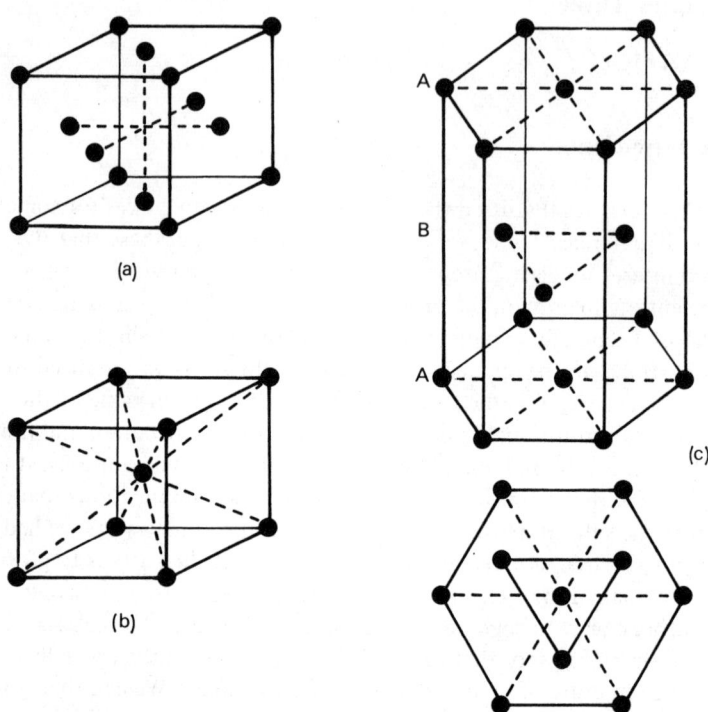

Figure 3.1   Cells showing (a) face-centred cubic, (b) body-centred cubic, and (c) close-packed hexagonal arrangements of atoms

The body-centred cubic and face-centred cubic arrangements are similar in that their unit cells are in the shape of a cube and that they have an atom located at each corner of the cube. The body-centred form, however, has another atom at the centre of the cube, whilst the face-centred form has an atom at the centre of each cube face. In the close-packed hexagonal form the atoms are located at the corners of a hexagonal prism and also at certain points within it, as shown in the figure. Both the face-centred cubic and the close-packed hexagonal arrangements have the important common feature — that of close-packing. They are the arrangements obtained by packing hard spherical objects together. For example, if spheres were packed together in a plane, they would form a layer marked A in Figure 3.2, corresponding to the basal plane A in Figure 3.1(c). The next layer of spheres would form a similar layer, marked B in the figures, but displaced so that the

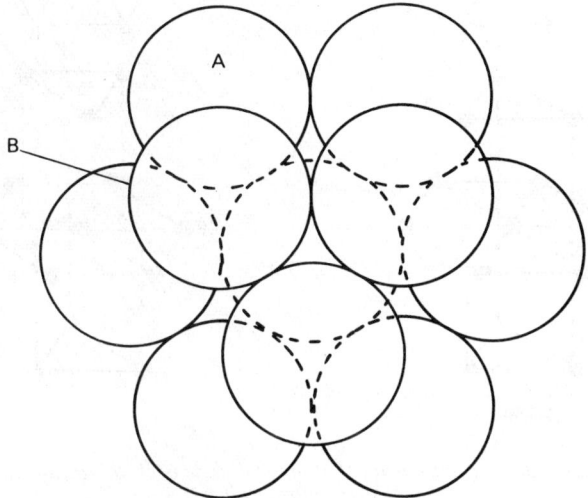

Figure 3.2.   Close-packed arrangement of spheres

spheres would rest in the depressions formed between three spheres of the layer below. The spheres of the third layer again fit into the depressions of the layer below but now there are two alternatives. The spheres could lie directly over those of the first layer, the basal plane, as in Figure 3.3(a), or they could be displaced again from those of the two lower layers so as to lie above the interstices of the A and B planes, as in Figure 3.3(b). In the first arrangement, with the third layer directly over the first, the stacking of the layers may be described as ABAB . . . and is the close-packed hexagonal arrangement, whereas in the second alternative the layers are arranged as ABCABC . . . . This second alternative is in fact another way of describing the face-centred cubic arrangement, as can be seen from Figure 3.4.

An alternative way of arranging the spheres in layers is shown in Figure 3.5, with each sphere having four closest neighbours in its own plane instead of the six as with a hexagonal arrangement. The spheres of the next layer would then fit into the depressions of the first layer, with the spheres of the third layer lying directly over those of the first

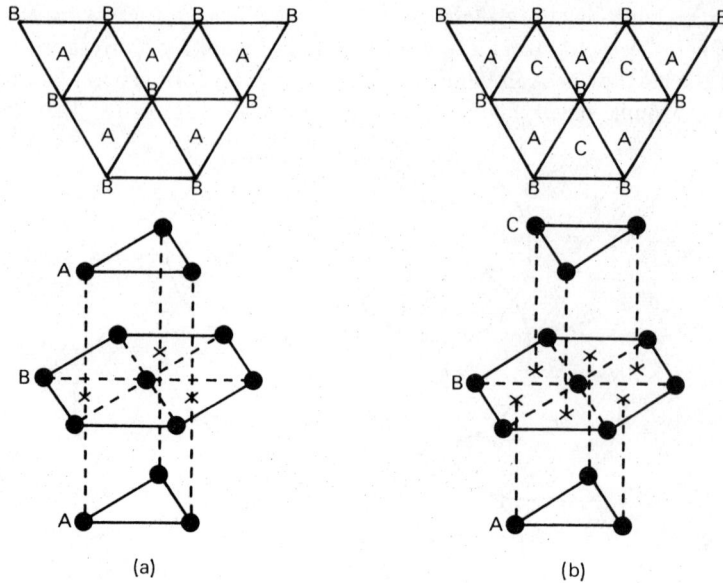

Figure 3.3. Arrangements of atoms: (a) close-packed hexagonal, and (b) face-centred cubic

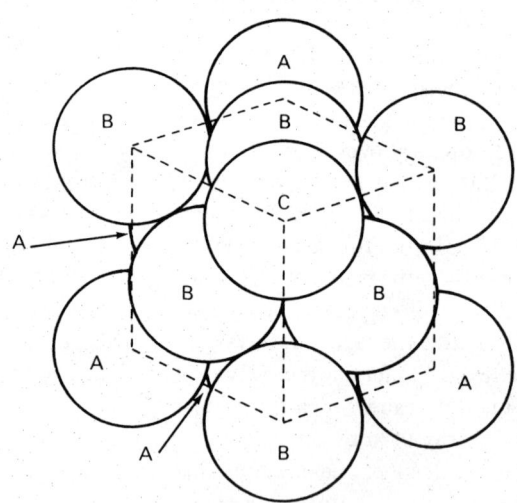

Figure 3.4. Face-centred cubic arrangement showing the ABCABC . . . sequence of planes

layer. This is the body-centred cubic arrangement and quite obviously
the spheres are not so closely packed as in the face-centred cubic or
hexagonal structures.

In these three types of structure there are certain particular layers
of atoms in which the atoms are especially close together and can be
imagined as forming slightly irregular plane surfaces which allow the

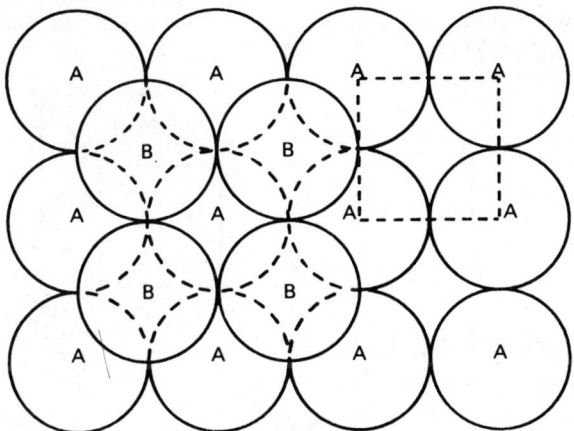

Figure 3.5.   Body-centred cubic stacking of layers of atoms

next layers of atoms to slide over them with comparative ease. To be
able to refer to such particular planes as these, and also to directions
within the crystal lattice, it is necessary to resort to a method of
labelling — the Miller notation.

To represent a crystallographic plane or one parallel to it, the plane
is imagined extended to cut the x, y, and z axes defining the edges of
the unit cell, as shown in Figure 3.6. The intercepts of the plane with
these axes are measured in terms of units of a, b, and c, the unit cell
dimensions. Reciprocals of these intercepts are then taken and the
fractions cleared to give the smallest integers. For example, if in
Figure 3.6 a, b, and c define the dimensions of the unit cell, measured
along the axes x, y, and z as shown, then the plane ABC may be
described by the Miller notation as follows. The intercepts OA, OB, and
OC on the x, y, and z axes respectively are 1, 2, and 2 measured in
units of a, b, and c respectively. The reciprocals of these intercepts are
$1, \frac{1}{2},$ and $\frac{1}{2}$, and multiplying through to remove the fractions we obtain
the Miller indices (211) of the plane ABC. Because of the repetitive

93

nature of the structure, any plane parallel to this will be described by the same indices. Negative intercepts are indicated by means of a bar over the appropriate index. A plane parallel to, say, the xz plane and intercepting the y axis at two units would have intercepts on the three axes ∞, 2, and ∞. Taking reciprocals gives 0, $\frac{1}{2}$, and 0 and multiplying through to obtain the lowest set of integers gives the Miller indices (010) for the plane. It will be seen that the plane of atoms marked B in Figure 3.4 is a close-packed plane and, by comparison with Figure 3.6, it has the indices (111).

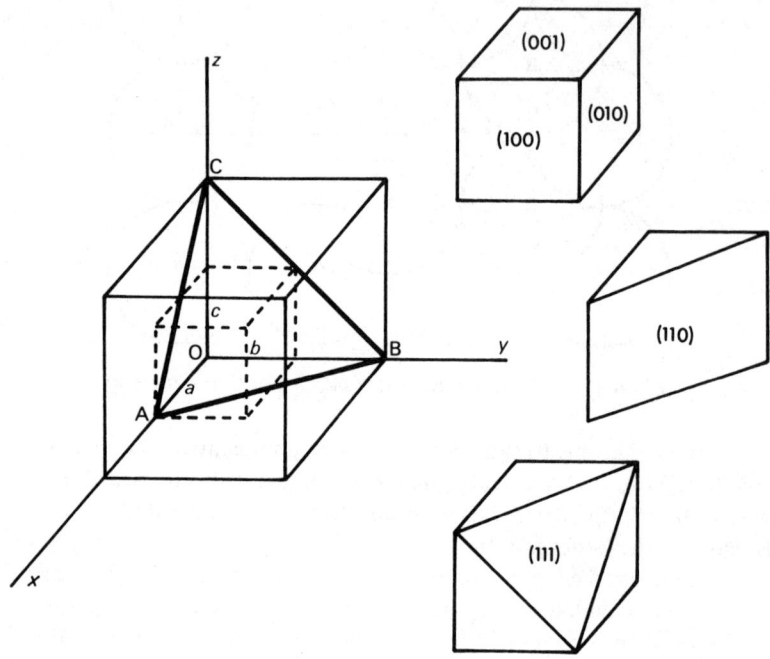

Figure 3.6. Miller indices of various crystallographic planes

A direction within the crystal may be indicated by a similar method of notation. The direction is first described in terms of its vector components resolved along each of the coordinate axes x, y, and z. These components are again measured in terms of the dimensions a, b, and c of the unit cell. For example, the direction of the line OA in Figure 3.7 is described by the vector components $\frac{3}{2}$, 2, and 1 measured along the x, y, and z axes and in units of a, b, and c respectively. Multiplying through to obtain the smallest integral coefficients, we obtain the Miller direction indices [342]. Again a negative direction may be indicated by a bar over the appropriate index. Also as before, owing to the repetitive structure of the crystal, this is the direction of any other line parallel to OA. The use of the different shaped brackets should be

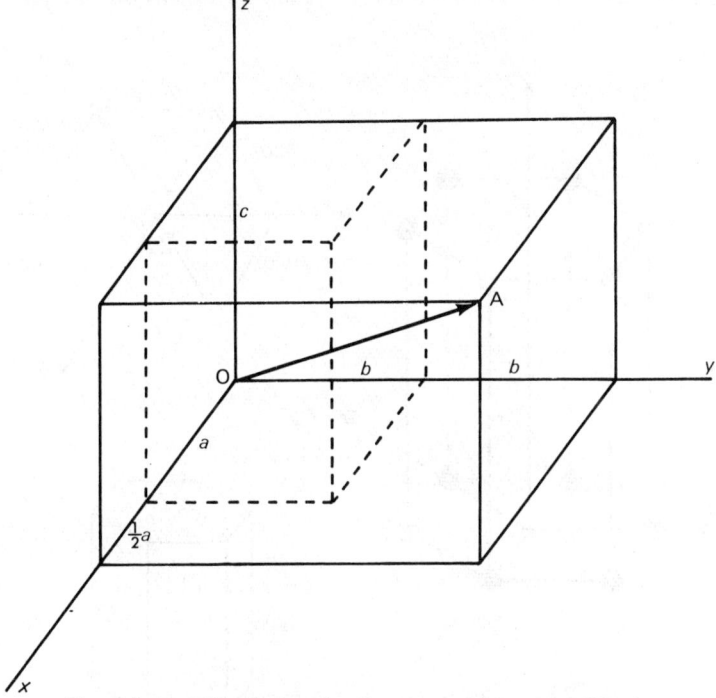

Figure 3.7. Miller indices for directions within a crystal lattice

noted — (*hkl*) indicates a plane and [*uvw*] indicates a direction. The notation {*hkl*} may be used to indicate the whole set of planes (*hkl*), (*khl*), ($\bar{h}kl$), etc. These are not parallel planes but they belong to the same family. In a cubic crystal all the planes in the same family have identical arrangements of atoms and consequently the same character. Similarly, the notation <*uvw*> is used to indicate a set of directions [*uwv*], [*wuv*], etc. For a cubic system it will be seen that the direction [*hkl*] is perpendicular to the plane (*hkl*).

With a slight modification to the system, planes and directions within a hexagonal array may be named. For example, the plane ABCDEF of the hexagonal cell in Figure 3.8 makes intercepts ∞, ∞, ∞, and 1 on the axes x_1, x_2, x_3, and z respectively. This plane is therefore described by the indices (0001). Similarly, the other faces are described by the indices shown in the figure.

Although we are concentrating mainly on the more common engineering metals because they are of chief interest, we must not forget some other materials that have importance in their own fields, for example

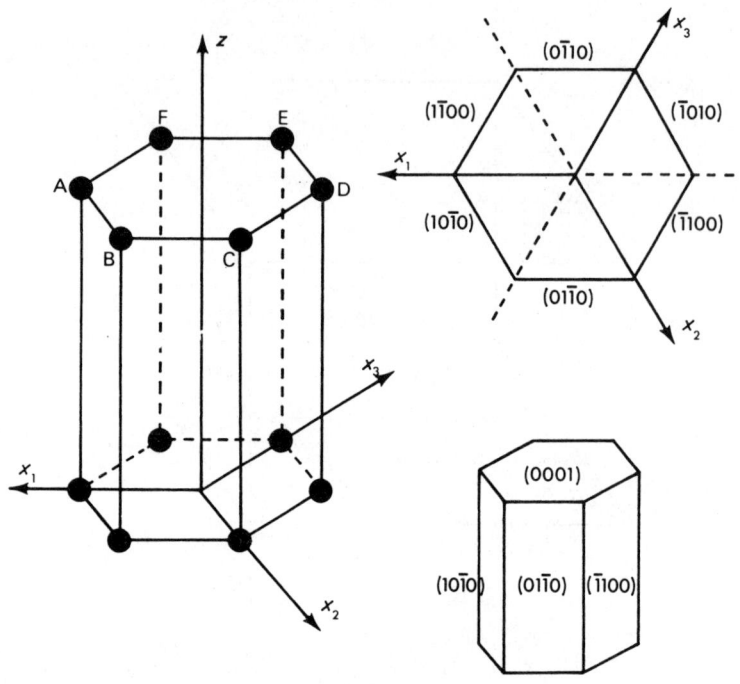

Figure 3.8. Miller–Bravais indices of a hexagonal structure

the polymers. A polymer is a material made up of very large molecules within which a particular atomic group or arrangement is repeated very many times. In fibres, long chain molecules are arranged with their axes substantially parallel to the fibre axis. For example, polyethylene molecules comprise repeat units of —CH_2—, perhaps many thousands of repeat units within one molecule. Long chain molecules may give crystalline or non-crystalline materials depending on their ordering. The molecules may be folded in a zig-zag manner giving ordering in only a small volume. Many polymer materials also contain spherulites. These

96

consist of many crystals growing out from one point, so a certain crystallographic direction is always towards that point.

Alternatively, a polymer may be a material in which there is a network of atoms joined by primary bonds as in the inorganic glass structures. There is a spatial repetition of units rather than a linear one as occures in the long chain molecules. For example, silica glass is made up of tetrahedra, each with four divalent oxygen ions at the corners and a quadrivalent silicon ion at the body-centre position. These tetrahedra are not arranged in a regular array but are arranged irregularly with the tetrahedra sharing corners. Even in the molten state the tetrahedra still arrange themselves so as to share corners and in this sense there is no difference in the type of arrangement between the liquid and the solid states. Glass may thus be termed a 'supercooled' liquid. It is an amorphous material, that is, it is not crystalline and it has no long-range ordering.

Depending on the nature of the bonds between molecules in a polymer, it may or may not soften on heating. In a long chain polymer the side bonds between molecules may be very weak. Thus, on heating, the bonds are easily broken allowing the molecules to slide over each other. This would be termed a *thermoplastic* material, as would silica glass. Some polymers have strong bonds between molecules, as does cellulose. It has long chain molecules but with numerous strong hydrogen bonds between the chains and, on heating, the material decomposes before the side bonds are broken. Thus it does not soften and is termed a *thermosetting* material.

3.3 Deformation Mechanisms

If a body is deformed elastically, that is, it is strained by only a small amount, the interatomic bond lengths will be changed by a small amount. The work done by external forces in changing the lengths of these bonds is stored in the bonds so that, if the external stress is removed, this bond energy will return the external work done. The elastic deformation process is thus reversible.

When a polymer is deformed, other processes may operate. For example, rubber is made up of long chain molecules that are very kinked and intertwined. Owing to thermal vibrations these entangled molecules are constantly being shuffled in a random manner. As the rubber is strained, there is a tendency to straighten out the molecules

97

and reduce their randomness, so the external work now goes to reduce the effect of the thermal vibrations. The entropy is thus reduced. On release of the external stress, the rubber returns to its original shape owing to the thermal agitation randomising the structure again. It is then in its original condition.

If now a body is deformed by a greater amount, the bonds between neighbouring atoms may be broken, atoms will move to new relative positions, and fresh bonds will be made with their new neighbours. On release of the external stress, the atoms will stay in their new positions, since these new bonds are of the same form and hence as good as the original ones. The body is thus permanently deformed – the deformation is an inelastic one. If new bonds are not formed for some reason, perhaps because of too great a stress or a high stress concentration due to a crack, then the body will fracture.

In practice it is found that this inelastic deformation may take place with stresses far lower than that expected from a consideration of the strength of the atomic bonds. In an amorphous polymer such as glass the thermal vibrations of the lattice become very important. Here there is a breaking of bonds in localised areas leading to a localised movement of atoms and molecules, without any effect on those farther away. This process will be going on throughout the bulk of material so, if now an external stress is applied, its action will be to bias the breaking of bonds and movement of atoms in the direction of the stress. The body will thus deform and, if the biasing is sufficiently great, there will be a deformation in which the strain rate is proportional to the applied stress. In other words there is a *viscous* flow of material, and it is the effect of thermal energy that makes this flow possible. If the temperature is low, there may not be sufficient thermal energy to break the bonds. There will thus be no flow until the external stress is large enough to do the breaking of the bonds, and this usually leads to brittle fracture.

In a crystal the mechanism of deformation is different again. In this case a deformed region, due to some imperfection in the lattice, is caused to move through the lattice by a relatively small externally applied stress. Thermal vibrations need play no part in this process since there can be no large shift of atoms in a localised position as there could with an amorphous material. If there were, the material would no longer be crystalline with its associated regular lattice arrangement. The energy required to break a set of bonds is now obtained mainly from the energy released by the joining of the preceding set of bonds, so the energy is passed along with the moving deformation. The external stress need only be sufficient to give impetus to this movement and bias its direction of motion. This movement of a deformation may be likened to the movement of a ruck in a large carpet. If the edge of a carpet is held and it is moved smartly up and down, a ruck may be produced that will travel the length of the carpet. If the carpet is pulled along at the same time, it will be found to slide along with comparative ease as the ruck moves across.

It is the causes and types of these imperfections and deformations within a crystal and their subsequent movements that constitute the subject matter of most of the remainder of this chapter, since it is these deformations that affect so dramatically the properties of the materials of our main field of interest, the metals and their alloys.

3.4 Lattice Defects

When a metal is molten, its atoms have no long-range order. There may be some short-range ordering but, owing to thermal vibrations, this is continuously being changed in a random manner. When the metal freezes, the solidification will take place in stages until the whole becomes solid.

The first stage in the freezing process is the formation of a number of solid-phase nuclei which grow in size as the metal cools. The number of these nuclei will depend on the impurities present, if any, and the rate of cooling. Increased rates of cooling and clusters of impurity atoms generally produce increased numbers of nuclei. As each solid-phase nucleus grows at the expense of the surrounding liquid, it forms into a regular crystalline array of atoms. At first the growth may be rather uneven, termed *dendritic* growth, but it will normally soon settle down to form a regular crystal. If the whole specimen cools down uniformly without any temperature gradients, then the crystallographic directions within each solid-phase nucleus will be only randomly orientated to those of its neighbours. As these nuclei grow, they will become grains that will grow until virtually the whole of the metal is solid. Then the atoms of the small amount of liquid left between the grains will have insufficient room to join onto either of the regular lattices of the neighbouring grains. These atoms can form only an irregular arrangement in the grain boundary region, which is usually about two to five atomic layers thick. The metal is then all solid. A polished section of the metal, etched to show up the grain boundaries, may then have the appearance of Figure 3.9. The crystallographic lattices of the individual grains are randomly orientated and the grain boundary regions may contain many of the impurities – those atoms that could not fit into the metal lattice. Certain grains, however, may come together at a boundary with a special orientation, as shown in Figure 3.10, to form a *twin* boundary; across this boundary the lattice forms a mirror image

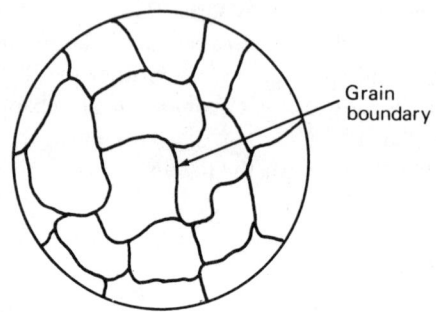

Grain
boundary

Figure 3.9.   Diagrammatic representation of a section of solid metal showing the grains

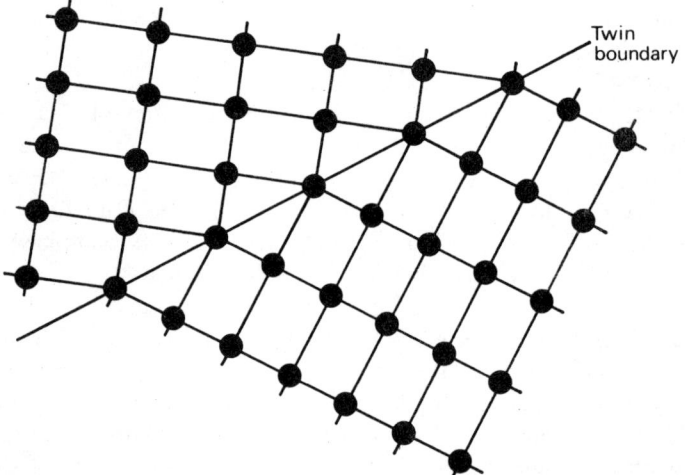

Twin
boundary

Figure 3.10.   Representation of a twin boundary

of itself. Each grain, however, will not be a perfect lattice arrangement. It may have a localised fault (a point defect), or a fault involving a row of many atoms (a line defect), or it may have a planar defect which involves whole planes of atoms. Generally a grain will most likely have all these types of defects together.

A *point defect*, as its name implies, is a very localised disorder and can take various forms. When an alloying element is deliberately added or, what comes to the same thing, an impurity is present, a foreign atom may try to fit into the lattice in place of a host atom or into a space between the host atoms. These are termed substitutional and interstitial atoms respectively and are illustrated in Figure 3.11. They are *impurity*

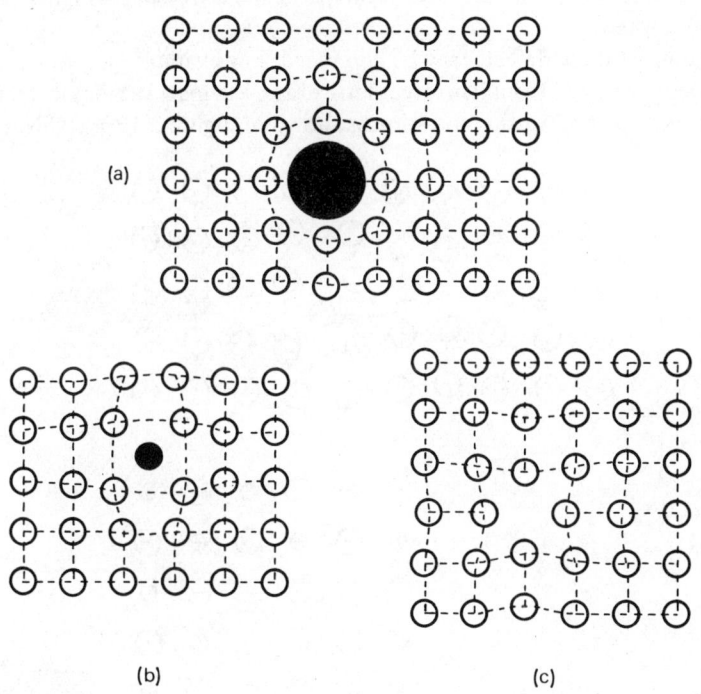

Figure 3.11. *Deformation of a crystal lattice due to (a) a substitutional atom, (b) an interstitial atom, and (c) a vacancy*

defects and cause distortion of the lattice in their vicinity. If a substitutional atom is similar in size to the atom it replaces, then this distortion may be quite small. Too great a misfit would limit the number of these impurity atoms that could be squeezed into the lattice. If an atom is missing from a lattice site, then a *vacancy* occurs, again causing a deformation in a small region as in Figure 3.11(c). Of course, if a number of vacancies should come together a *void* would be formed. This will tend to happen as the temperature is raised since there is more chance of a neighbouring atom moving into a vacancy, thus causing the vacancy to move one atomic position at a time. Also, the excess energy within the crystal will be reduced if such point defects coalesce. In the same way that vacancies become mobile with increasing temperature, so also do interstitial atoms. Again the excess energy in the crystal may be reduced if they coalesce, and this time they will form a *precipitate*. If large enough, they may constitute a new phase; a different phase can be a different material or the same material in a different form, like ice and water.

Some of these defects may be combined. For example, an atom may become displaced from its correct lattice site, leaving a vacancy defect, and take up an interstitial position as shown in Figure 3.12(a). This is

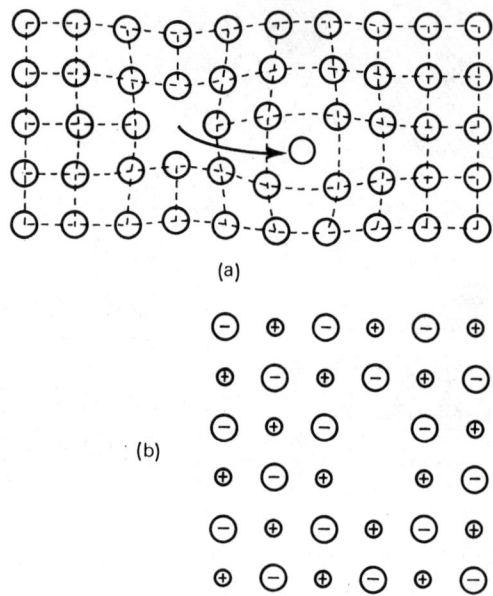

(a)

(b)

Figure 3.12. (a) A Frenkel defect, and (b) a Schottky defect

termed a *Frenkel defect*. In an ionic solid a vacancy due to the absence of a positive ion may be balanced by a vacancy due to the absence of a negative ion. This is termed a *Schottky defect* and is illustrated in Figure 3.12(b).

A *line defect* is characterised by a fault along a row of atoms and is termed a dislocation, which may be one of two types. An *edge dislocation* is shown diagrammatically in Figure 3.13. The dislocation runs normal to the page in Figure 3.13(a) and is indicated by the symbol ⊥. An extra row of atoms occurs above the dislocation, so the crystal above is under compression whilst that below the dislocation is under tension. An edge dislocation may be considered to be the line boundary within the crystal of an extra plane of atoms.

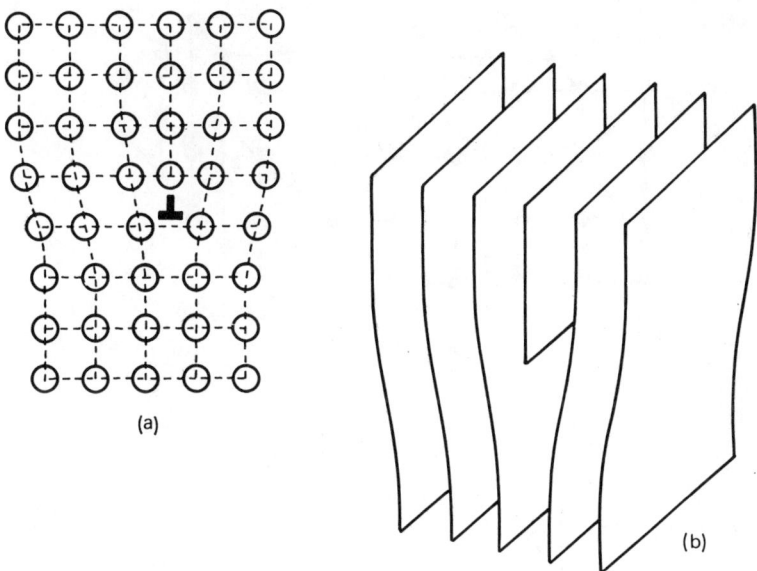

(a)

(b)

Figure 3.13. An edge dislocation: (a) occurring in a lattice, and (b) shown diagrammatically

In a *screw dislocation* a row of atoms has its right number of nearest neighbours but their orientation is distorted, as shown in Figure 3.14. If a perfect crystal is imagined partially cut through in a plane parallel to a principal plane of atoms and then the parts of the crystal on either side of the cut are sheared relative to one another by an amount equal to the atomic spacing, the atomic bonds may again be correctly made. A screw dislocation has then been introduced along the line AB, parallel to the direction of shear, as in the figure.

An alternative way of visualising edge and screw dislocations is to consider the crystal to be partially sheared, as shown in Figure 3.15. If part of the crystal, down to the line AB, has been sheared one atomic spacing in the direction shown in Figure 3.15(a) so that part of the crystal has moved along a slip plane, then AB will be an edge dislocation.

103

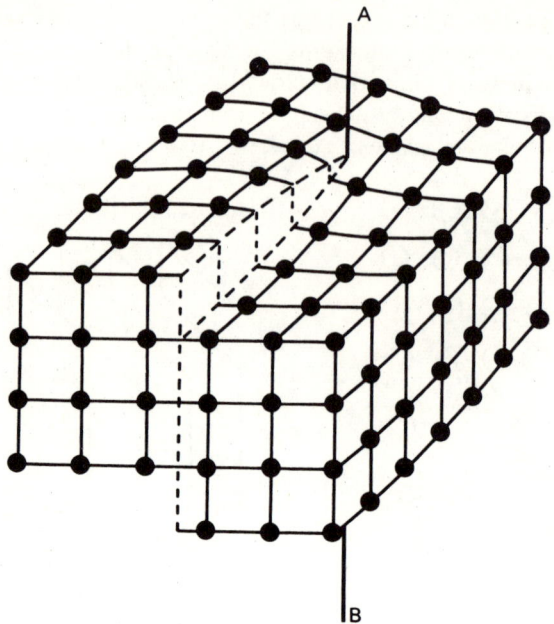

Figure 3.14. A screw dislocation

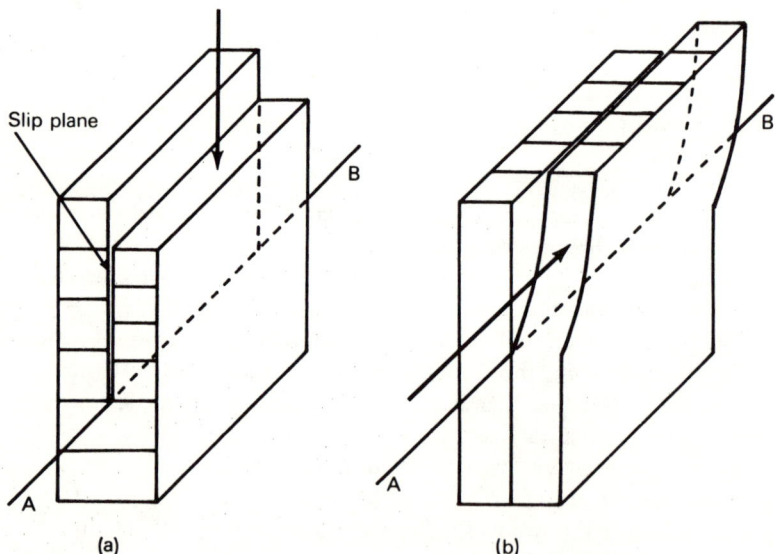

Slip plane

(a)   (b)

Figure 3.15. Representation of (a) an edge dislocation, and (b) a screw dislocation

That is, the slip direction is perpendicular to the dislocation line. If, however, part of the crystal had been sheared in the direction shown in Figure 3.15(b), then a screw dislocation would have been formed. Now the slip direction is parallel to the dislocation line AB.

Bigger shears than one atomic spacing are possible. In practice both edge and screw dislocations may be combined and may blend from one to the other as a dislocation meanders through the crystal lattice. In fact a pure edge or screw dislocation alone is comparatively rare.

What evidence is there that dislocations exist? Dislocations are regions of high energy due to the localised strain produced so, by using chemical etchants, etch pits are formed where the dislocation sites are preferentially attacked. This method can be very effective on polished or cleavage faces provided the density of dislocation is fairly low. Direct observation of dislocations is possible by electron microscopy techniques where the lattice spacing is sufficiently large, as in some organic crystals. For metals the lattice spacing is too small in comparison with the resolution of an electron microscope and direct observation of the individual atoms and planes is not possible. However, by such techniques as the scattering of the electron beam in transmission electron microscopy to obtain contrast in the image and the use of Moiré fringes, the presence and movement of dislocations in metals has been studied. The Moiré fringe technique makes use of the overlap of two lattices of slightly different spacing or orientation which in combination give an image that appears to be of a lattice spacing greater than either of them. A dislocation in one of the lattices will give a modification to the resultant Moiré pattern.

The strength of a dislocation, a measure of the amount of distortion introduced into the lattice, is measured in terms of the *Burgers vector*. Considering first an edge dislocation and with reference to Figure 3.16(a), the Burgers vector may be found as follows. A circuit is drawn through the lattice in a plane perpendicular to the line of the edge dislocation, with the dislocation approximately at the centre of the circuit. For example, starting at the point A in the figure, count x atomic units to B, past the dislocation, and then y atomic units to C, again past the dislocation, x units across to D and finally y units up to E. If there were no dislocation, a complete circuit would have been made so that E would have coincided with the starting point A. In our case the distance EA required for the closure of the circuit is a measure of the amount and consequently the strength of the dislocation. The vector joining the end of the circuit at E to the starting point A is termed the Burgers vector. It will thus be seen from a consideration of the figure that the Burgers vector for an edge dislocation is always perpendicular to the line of the dislocation. When drawing the Burgers circuit, care must be taken that the circuit is not too close to the very distorted region immediately surrounding the dislocation.

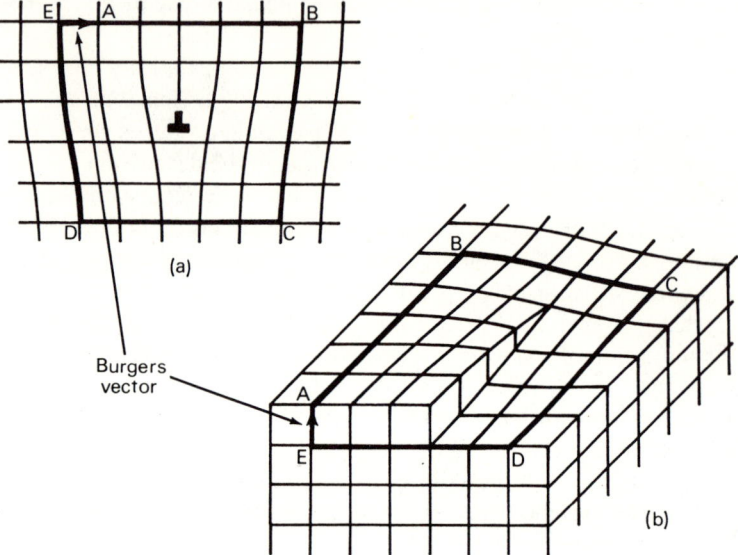

Figure 3.16. A Burgers circuit and Burgers vector drawn for (a) an edge dislocation, and (b) a screw dislocation

In a similar manner the strength of a screw dislocation may be measured by drawing a Burgers circuit in a plane nominally perpendicular to the axis of the screw, as shown in Figure 3.16(b). The closing vector EA is the Burgers vector, and it will be seen from the figure that it is always parallel to the dislocation line. If the Burgers circuit is continued to be drawn, it will trace out a spiral path whose pitch is equal to the Burgers vector.

The Burgers vector is a very important property of a dislocation. It is a constant. No matter how the dislocation may wind through a crystal, it does not change its Burgers vector. It follows then that a dislocation cannot just terminate within a crystal. It can only terminate at a free surface, which could be an external surface of the crystal or the surface of a void within the crystal, or it may terminate in the sense that it meets its other end to form a continuous loop, or it may join with other dislocations at a node.

3.5 The Movement of Dislocations in Crystals

As was said in Section 3.3, it is the motion of the deformations and imperfections within a crystal lattice that so strongly affect the non-elastic properties of a crystal structure — in our case, that of a metal. Therefore we must now discuss in detail how these dislocations move, and their effect.

A dislocation may have any orientation within a crystal but, owing to its very nature, a dislocation will generally be on one of a number of definite crystallographic planes — usually planes of densest atomic packing. Let us consider first the movement of an edge dislocation in such a densely packed plane. Figure 3.17(b) shows a cross-section of part of

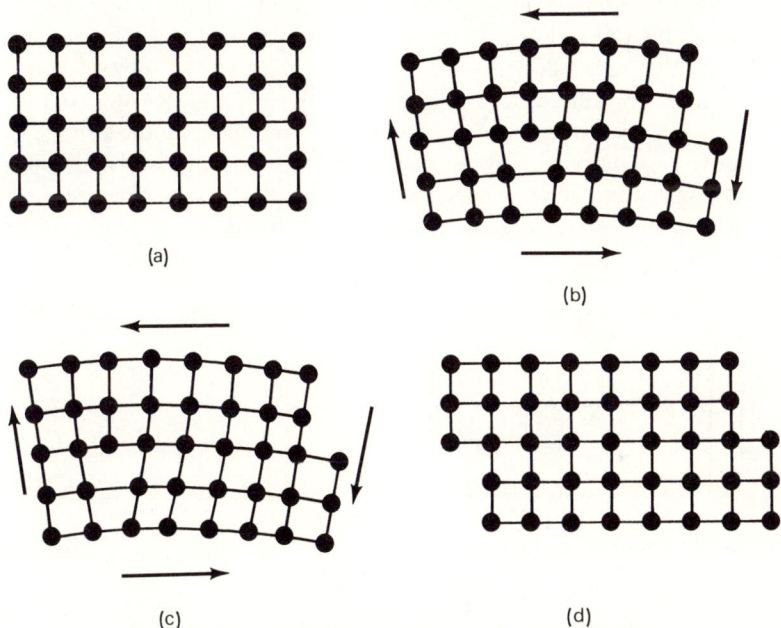

(a)

(b)

(c)

(d)

Figure 3.17. Showing the progressive movement of an edge dislocation across a crystal: a perfect crystal (a) is deformed to a sheared dislocation-free crystal (d) by the movement of a dislocation (b), (c)

a crystal with an edge dislocation normal to this plane, that is, we are viewing the end of an extra half-plane. With a shear stress applied as shown, the extra half-plane can be shifted across the crystal as in Figure 3.17(c) until it reaches the surface of the crystal. It then forms

107

a slip step. Slip has now been propagated through the crystal in stages, atomic layer by atomic layer – it has not occurred by the simultaneous movement of all the atoms along the slip plane and consequently only a small shear stress was required to produce this plastic deformation. This type of dislocation is also known as a *glide dislocation*. It has its Burgers vector in the direction of the slip plane and, as its name implies, it allows a slip to glide through with comparative ease and also without any diffusion of material through the lattice. The process is also said to be a conservative one because the nature of the planes, especially the extra half-plane of atoms, is not altered by the glide.

A similar effect can be achieved with two such edge dislocations, but of opposite sign. A sign may be arbitrarily allocated by considering an edge dislocation formed by the addition of the extra half-layer of atoms to the top half of the crystal to be a positive edge dislocation, whilst if the extra half-plane occurs in the bottom half of the crystal it is negative. The edge dislocations drawn in Figures 3.13, 3.16(a), and 3.17 are all positive and are conventionally symbolised by the sign ⊥. A negative edge dislocation is symbolised by ⊤. If now a crystal containing both positive and negative edge dislocations with the same Burgers vector is sheared as shown in Figure 3.18, a dislocation-free sheared crystal will

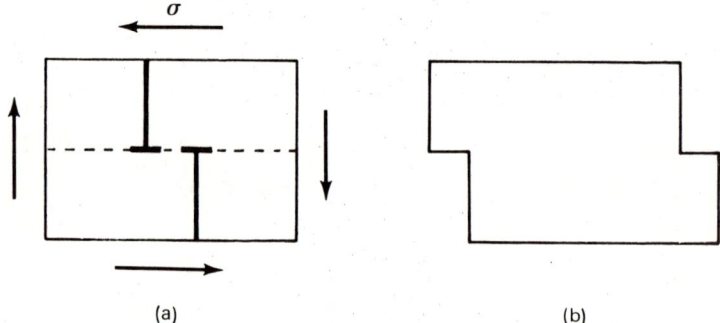

(a)   (b)

Figure 3.18. A deformed dislocation-free crystal (b), formed by the movement of edge dislocations of opposite signs as in (a)

be formed as before. If instead the shear stress had been applied in the opposite direction, the dislocations would have moved towards each other and, on meeting, would have formed a complete layer of atoms. The dislocations would have annihilated each other, leaving an undeformed dislocation-free crystal.

If a screw dislocation is moved right through the crystal by an applied stress, it will likewise produce a dislocation-free crystal with a

108

slip step. The net effect will thus be the same as if an edge dislocation had travelled through. The motion of a screw dislocation under an applied stress is shown in Figure 3.19. In the same way that two edge

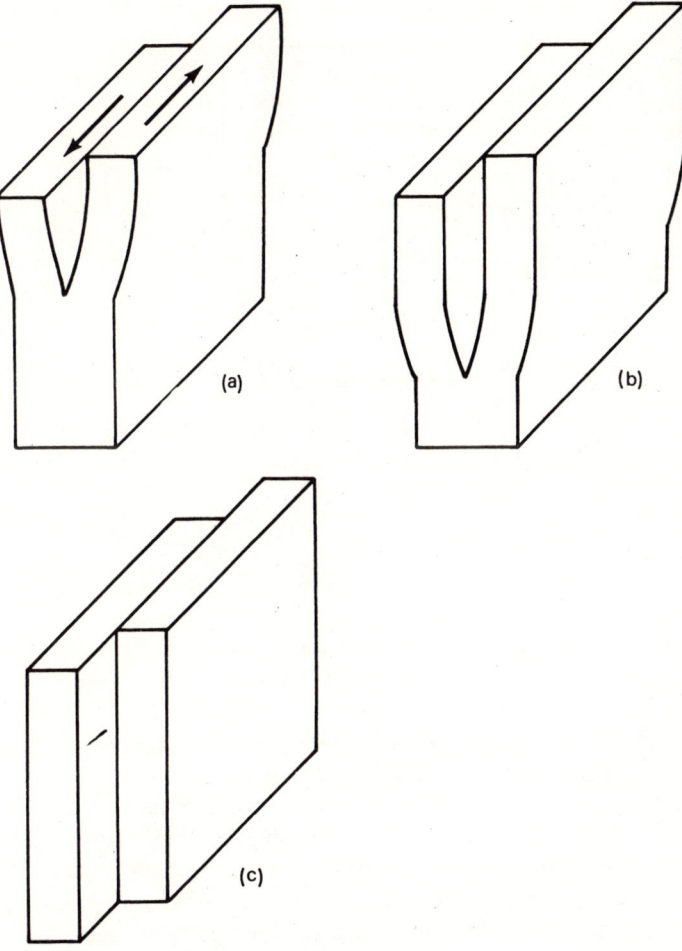

Figure 3.19. Motion of a screw dislocation across a crystal under the action of an applied stress

dislocations of opposite sign can annihilate each other on meeting, so also can two screw dislocations of opposite sign and the same Burgers vector.

109

It may be seen from the foregoing that the movement of an edge dislocation is normally restricted to a movement across a fixed slip plane. A screw dislocation, however, does not have this restriction. It can *cross-slip* so as to move in a slip plane parallel to the original slip plane, as shown in Figure 3.20. For an edge dislocation to shift onto

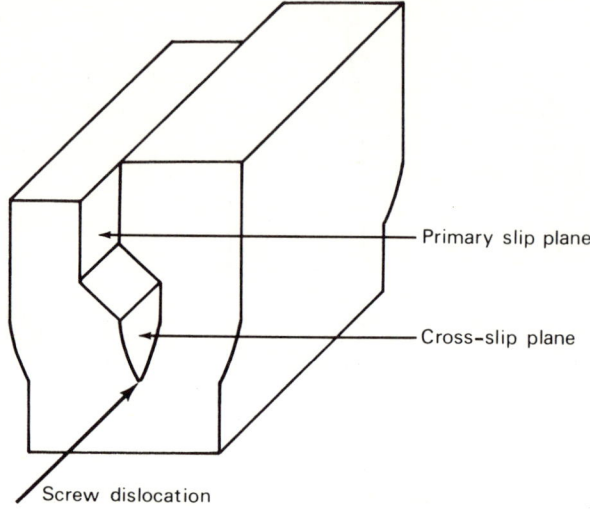

Figure 3.20.   Cross-slip in a screw dislocation

another slip plane, it is necessary for some diffusion of material to take place. This can happen, for example, if a vacancy diffuses up to an edge dislocation. The half-plane of atoms will shorten and the dislocation will *climb*. In general this will not happen all along the dislocation as the vacancies will arrive in small numbers at a time. The whole length of the dislocation cannot then all climb together and a *jog* will occur, as shown diagrammatically in Figure 3.21. If extra atoms, for example interstitial atoms of similar size to the host atoms, migrate to the dislocation, the jog will be in the opposite direction. Owing to the fact that a diffusion of material is required for climb to occur, it does not occur unless the temperature is sufficiently high, generally above about half the absolute melting temperature.

So far we have only discussed edge and screw dislocations and have seen how they may move across the plane in which they lie. More complicated movements are, however, possible. If a number of vacancies come together on a slip plane, a dislocation loop will be

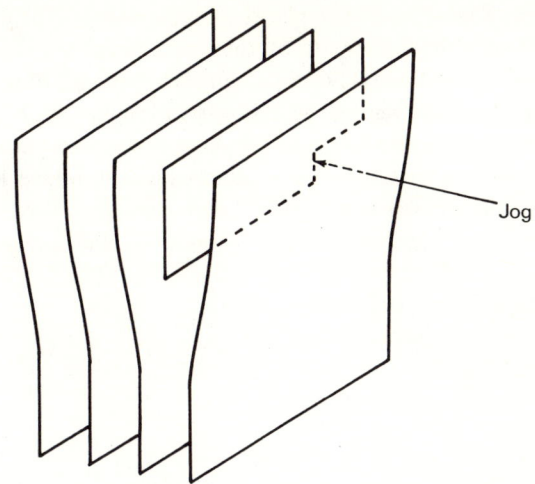

Figure 3.21. Diagrammatic representation of a jog in an edge dislocation

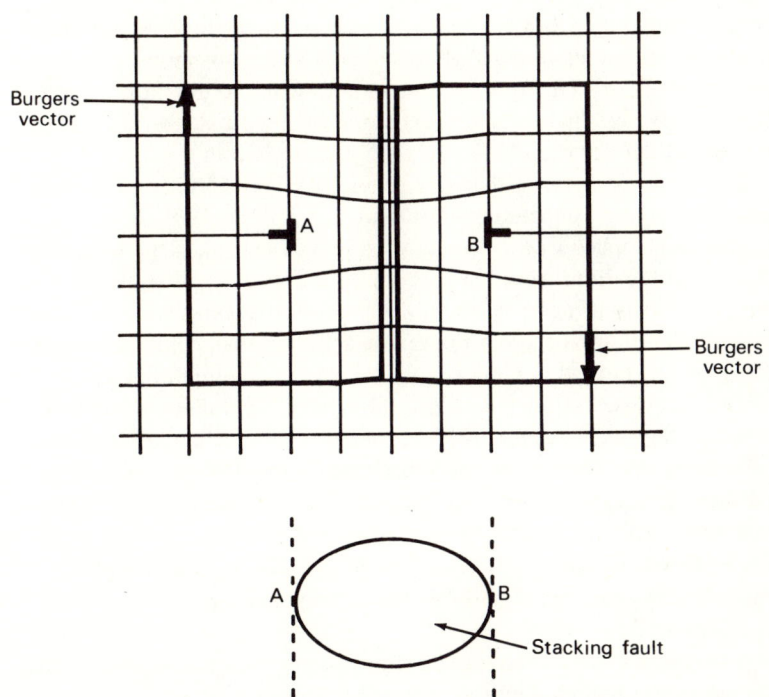

Figure 3.22. Formation of a dislocation loop

formed as in Figure 3.22. The planes on either side of the vacancy cluster will partially collapse to fill the space, forming a stacking fault in this region since there will be an incorrect sequence of planes. As will be seen from the figure, this dislocation loop has its Burgers vector in a direction perpendicular to the plane of the loop. Such a dislocation loop is termed a *sessile* dislocation and it cannot move freely in the plane in which it lies. The only way it can move in its own plane is by dislocation climb, involving the migration of vacancies or interstitials. This type of movement of the sessile dislocation, due to a mass transport of material, is a *non-conservative* motion.

3.6 Dislocation Multiplication

Returning now to the problem of an explanation of the reasons for plastic flow, it has been shown that the motions of dislocations allow a crystal to be deformed by stresses that are less than those needed to break all the bonds. For example, it is not necessary to break simultaneously all the bonds across a whole plane of atoms to cause slip between two parts of a crystal. We have shown that, owing to the motions of dislocations, the slip can take place atom row by atom row to cause a permanent deformation in the crystalline material.

In an annealed crystal there are generally $10^2 - 10^4$ dislocations per square millimetre, depending on the state of perfection of the crystal. If a stress is applied to the crystal, we might expect from the foregoing that the dislocations would move across the crystal to allow its deformation, eventually using up all the dislocations. In fact this is not so since a far greater plastic deformation can be achieved than could come from this large but limited number of dislocations. It is found in practice, from direct observation and x-ray studies, that the number of dislocations actually increases by orders of magnitude as the crystal is plastically deformed. This implies that dislocations must be created in large numbers. It may be shown that thermal vibrations are not sufficient to create the fresh dislocations that are observed at the low stress conditions just sufficient to produce slip, and therefore a different mechanism must be envisaged. Frank and Read have proposed such sources of dislocations, as follows.

Consider part of a crystal to be slipped as in Figure 3.23(a) with the slipped region bounded by the edge dislocations at AB and BC. If now the edge AB is fixed, perhaps held by a line of impurities, then the edge

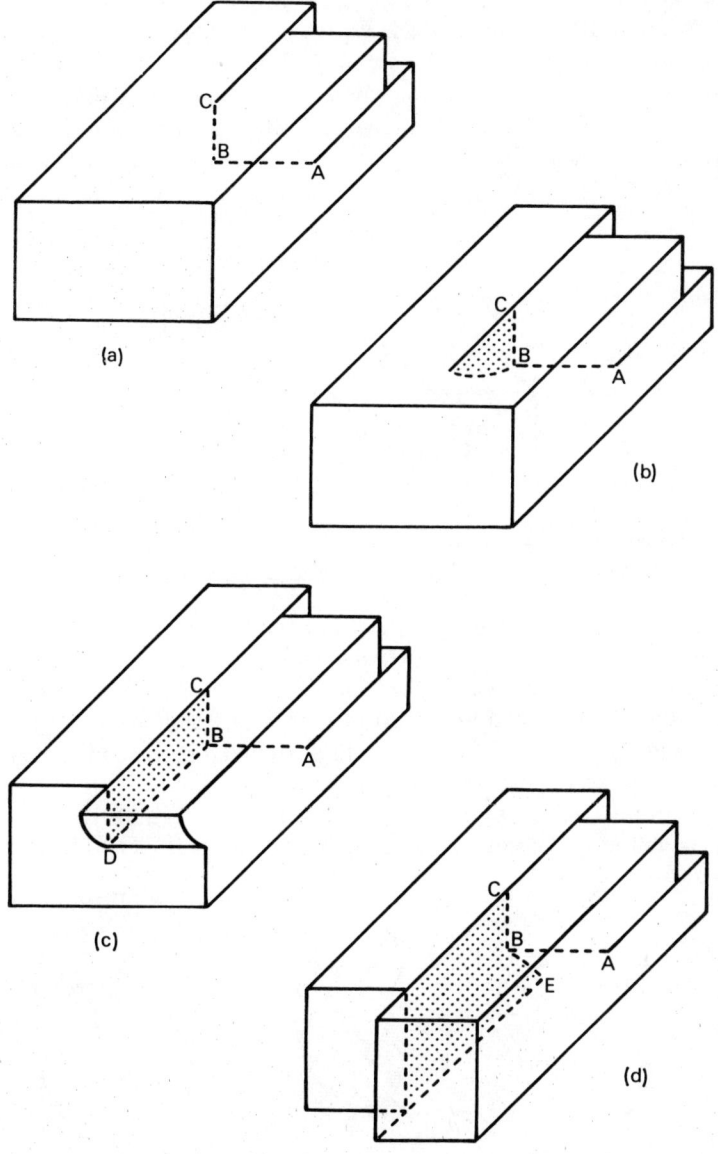

(a)

(b)

(c)

(d)

Figure 3.23. Motion of a Frank–Read spiral to produce slip over a complete plane

BC can only move by keeping the point B fixed, as in Figure 3.23(b). Continued movement of this dislocation produces almost a pure screw dislocation BD [Figure 3.23(c)]. Further movement of the dislocation, through the position BE of Figure 3.23(d), will eventually cause the slip of one atomic unit over the whole plane swept out by the dislocation, eventually rotating to its original position BC when the crystal will be deformed to the shape shown in Figure 3.24. This process can

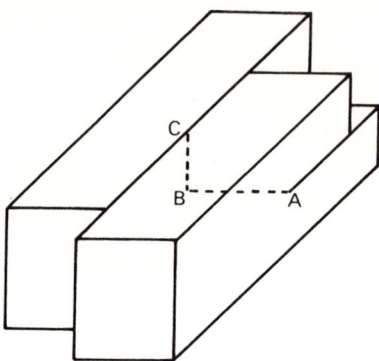

Figure 3.24.   Final shape of crystal deformed as in Figure 3.23

continue with each complete rotation of the edge BC about its anchor point B producing a unit slip over the whole plane.

An alternative mechanism for producing repeated slip over a plane is by the operation of a Frank—Read source. This is shown in various stages in Figure 3.25. Here an edge dislocation has become pinned at two points, again possibly by impurities or defects in the lattice. An applied stress will cause the dislocation to bow out from its straight line condition to produce an area, shown dotted, over which slip has

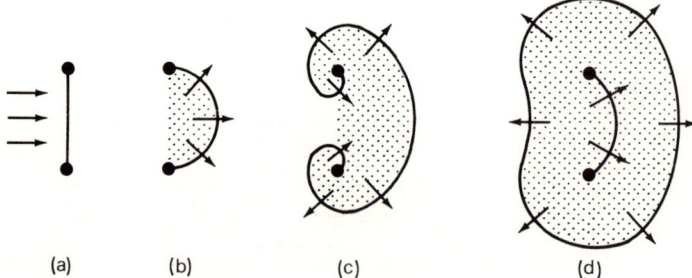

(a)   (b)   (c)   (d)

Figure 3.25.   Motion of a pinned dislocation to produce a Frank—Read source, allowing repeated slip to occur

114

occurred. When the dislocation edge becomes a semicircle it becomes unstable, and continuing stress will cause it to take the shape shown in Figure 3.25(c). The lobes will eventually touch and, being of opposite sign, will annihilate each other to form a closed dislocation loop. The remaining section carries through and is available for repeating the processes, as shown in Figure 3.25(d). Thus new dislocation loops may be continuously created, allowing repeated slip to occur in the plane.

3.7 Effect of Dislocations on Mechanical Properties

In Chapter 1 we were able to give some explanation for the cause of the elastic property of a material but, owing to the complexity of real materials and the number of possible variables involved in their structure, it was not possible to obtain numerical agreement with the observed elastic properties of real materials. In this chapter we have considered the structure of the materials of practical interest to us, mainly the engineering metals, and have seen that a practical material contains a vast number of imperfections. Apart from the granular nature of the materials, where each separate grain is a small crystallite with its lattice orientation generally at a random angle to its neighbour, the lattice of each crystallite is never perfect, containing point, line, and planar defects. The consequently vast number of variables, making every grain different from any other grain, again makes it virtually impossible to give any exact numerical agreement with the plastic properties of real materials. Nevertheless we are able to give a qualitative account which is in fair agreement with observed facts.

We have seen that the onset of plastic deformation, the yield strength, is mainly dependent on the movement of a large number of dislocations. It has been estimated that a perfect crystal, that is, one without any dislocations or other imperfections, would be about a thousand times stronger than a normal crystal of that material. We can here talk only of single crystals so as to avoid the great complexities introduced by the discontinuities at a grain boundary. This greater strength has been demonstrated to some extent by the properties of single crystal *whiskers*. These are about 10^{-3} mm in diameter and can be grown almost dislocation-free, although they must still be considered far from theoretical perfection. For iron whiskers, an increase in the ultimate tensile strength of about 30 times can be obtained over that for normal wrought iron.

The yield strength of a material may also be increased if the dislocations can be stopped or hindered from moving in some way. Thus large impurities that cause dislocations to pile up and stop them from continuing across the crystal will have this effect. This could be achieved by, for example, the technique known as *precipitation hardening.* An alloy may have a particular phase that becomes less soluble in another as the temperature is reduced. By quenching from a high to a slightly lower temperature, this phase will become supersaturated with respect to the other. If now the alloy is held at this temperature, the less soluble phase will be given time to precipitate out. The longer the alloy is held at this temperature, the larger the precipitated particles will grow. Cooling further to, say, room temperature will freeze the process, preventing any further growth of the precipitate. The dislocations interact with these regions of localised volume strain produced by the misfitting particles so that their motion is prevented or hindered, giving an increase in the yield strength of the material. The degree and nature of the interaction will depend on the size, shape, and distribution of the precipitated particles. These factors may be controlled by the type of alloying atoms used and the heat treatment conditions. Too large a particle can lead to weakness owing to possible poor bonding between it and the surrounding matrix, resulting in surfaces of weakness and possible fracture. For this reason small or even needle-like precipitates may be better. Hardening may also arise if the precipitate is such that the misfit between the lattice of the precipitated phase is only slightly different from that of the matrix. Both lattices will then be strained to give atom-to-atom correspondence over comparatively large regions. This internally strained condition has the effect of increasing the strength of the material in a manner somewhat analogous to that of prestressed concrete. This method of hardening is known as *coherence* hardening.

As a crystal is deformed and manipulated, that is, *cold worked*, for example by machining processes, the easy-to-move dislocations move first of all, followed in turn by the others, until they begin to be hindered in their motion. This hindrance will steadily increase in degree as the dislocations pile up behind impurities and are forced to take other routes, by moving in more difficult slip planes. As well as this, different dislocations will intersect and have to force their way through each other, with vacancies and interstitials being left behind at the intersections and so providing additional hindrance for following dislocations. Dislocations will also be combining to form sessile dislocations, leading

116

to a general build-up of increasingly more difficult to move dislocations. With this increased resistance to slip the material is said to be *work hardened*. New dislocations will also be produced by mechanisms discussed, for example, in the previous section, which in turn will add to the general pile-up. The number of dislocations may well increase from about 10^4 per square millimetre for an annealed specimen to 10^8 for a few percent deformation, increasing to perhaps 10^{12} per square millimetre after heavy cold working.

The grain boundaries also play a part in this as they too prevent dislocations moving very far. As has already been seen, slip will most easily occur in planes of close-packed atoms. If the direction of the applied stress does not lie in one of these planes, there will be a tendency for some of the crystalline grains to rotate slightly to get into a more favourable orientation. It is found that during this process some parts of the crystal have a tendency to lag behind. These regions are commonly in the form of platelets about 0·05 mm thick and occurring at about 1 mm intervals. They are orientated perpendicular to the slip plane, as shown in Figure 3.26, and run right across the crystal. They

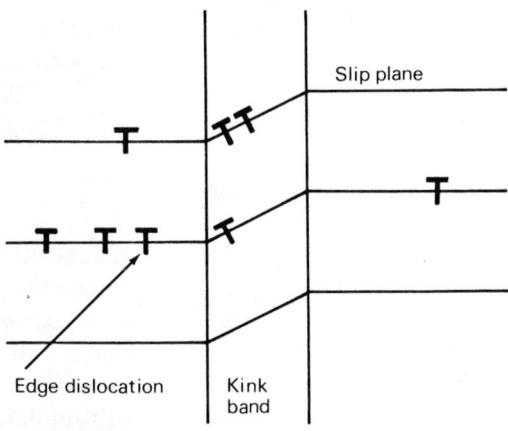

Figure 3.26. Rotation of a section of lattice to form a kink band

are termed *kink bands* and will cause an additional obstacle to the movement of a dislocation.

With increasing work hardening comes increasing possibility of fracture. If the material is unable to slip, then increased stress can only give the possibility of the rupturing of atomic bonds. If this happens in a small region, the stress will be increased over the remaining bonds owing to the reduction of the effective cross-sectional area, the broken bond region will rapidly propagate as a crack, and complete fracture will ensue. It may therefore be desirable to decrease the degree of work hardening. This may be especially necessary if further machining operations are required to be carried out — to speed the working and cutting operations and to reduce wear on the cutting tools. Work hardening may be reduced by an *annealing* process. In this the material is heated to some temperature higher than about one-third its absolute melting temperature and held at this value for a period depending on the temperature and amount of work hardening. Because of the work hardening, considerable strain energy may be stored within the crystal. By heating the crystal, diffusion of vacancies and other point defects can take place, allowing the dislocations to unlock themselves from these obstacles. The extra thermal energy also enables the dislocations to climb and cross-slip and generally to move across the crystal so as to reduce the number of dislocations and hence the locked-in strain energy. The material is then said to be annealed and is in a soft, easily deformed, state. Of course, any further deformation will reintroduce work hardening.

A metal specimen having been annealed, it may be too soft for its required purpose. By heating and then quenching to a lower temperature, precipitation hardening may be produced as already discussed. This is termed *tempering*. Its degree may be controlled by the particular heat treatment given, depending on the nature of the material and the type of precipitates formed.

Creep is also a process that may be explained by reference to dislocations. In Chapter 1 we saw that, if a constant stress is applied to a metal at an elevated temperature, the amount of strain increases with time. It can also increase at room temperature in, for example, lead. There is a rapid initial increase termed primary creep, followed by an extended period in which the strain increases at a constant rate, followed by tertiary creep when the strain rapidly increases with time and the specimen eventually fractures. The applied stress will cause the easy dislocations to move and the elevated temperature will give them sufficient energy to overcome small obstacles to their motion. Once this has happened, the primary creep condition changes over into secondary creep. Here the creep is at a steady rate and is due to the rate of increase of work hardening being balanced by the rate of recovery. This recovery rate will depend on the temperature since it is thermal energy that enables the piling-up dislocations to find other paths, mainly by climb onto other slip planes. In other words, the dislocations pile up by a work hardening process and then find a way round their obstacles to continue their passage and eventually produce slip, in this case creep. Thus the creep rate depends on the rate at which the dislocations are able to find a way around their obstacles, which in turn will depend on the thermal energy provided. At higher temperatures the grain boundaries play a part in the process. Then there is some sliding of grains past each other. The creep rate is here increased for a smaller grain size owing to the increased ease of sliding.

3.8 Microstructure

We have already said how the presence of dislocations may be revealed by etching and electron microscope techniques. Grosser effects, however, can be revealed directly by a microscope. By polishing an area of a specimen and etching with suitable chemical solutions, features such as grain boundaries, the relative orientation of the crystal lattice within individual grains, and precipitates are readily visible with an optical microscope, although for some types of precipitate the higher resolution of an electron microscope may be needed.

After the specimen has been deformed, a change in the grain shapes may be observed. Where part of the crystal has slipped on a slip plane, a surface step termed a *slip line* may be visible, although again with most metals the step may be below the resolution of an optical microscope and an electron microscope is necessary. Further deformation often leads to a clustering of slip lines to form a *slip band* which may well be visible with an optical microscope of low magnification. In kink bands, as we have already said, there is a rotation of one portion of the lattice within a parallel-sided layer (Figure 3.26) which takes place with respect to the surrounding lattice. This is usually clearly visible without optical magnification. Twinning may also occur within a parallel-sided region. A twin boundary is shown diagrammatically in Figure 3.10. A *deformation twin band* may again be visible by eye and will appear as a long parallel-sided band with a particular twin orientation to the rest of the crystal. *Lüders bands* may also be seen by eye in many deformed polycrystalline metal specimens. Owing to varying strain concentrations caused by the locking of dislocations in various regions in the early stages of the deformation, there may occur markedly softer regions. Deformation may then occur easily along the boundaries

119

of these regions producing long deformation bands as the material as a whole is deformed. Such bands may be many centimetres long. Finally, *flow lines* may be mentioned. As a metal is grossly deformed, the grains will be distorted and will tend to rotate so that their planes of easiest slip include the extension direction. Grains will become elongated and inclusions, that is, non-metallic impurities that have not dissolved into the matrix, will be drawn out into long lines termed flow lines.

3.9 Fibre Composite Materials

We have already mentioned that a material can have a much greater ultimate tensile strength when in the form of a crystalline whisker. Thus, by combining these whiskers side-by-side in a suitable matrix, a material of a useful cross-section combined with great tensile strength may be obtained. Not only whiskers but many other fibres and long filaments may be used. For example, one of the earliest composite materials was of glass fibre in a plastic matrix. A composite material has the advantage that greater strength coupled with a lower density relative to conventional structural materials can be achieved, with obvious benefits in, for example, the field of aeronautical engineering.

To produce a satisfactory composite, the fibres should be aligned parallel with each other and with the tension axis, otherwise shear forces will be introduced to which the fibres may be less resistant. They should not touch each other and should have the same strength, size, and shape. Care must be taken not to damage the fibres since a scratch, or even a small blemish, on the surface may act as a site from which a crack may propagate and hence considerably reduce the ultimate tensile strength.

The matrix material in its turn should be capable of wetting the fibre surface in the fabrication process so that, on setting, each fibre will be bonded to the matrix. Thus the fibres and the matrix will be elongated equally when a tensile stress is applied. As well as being chemically inert with respect to the fibres, the matrix material should have a lower modulus and be more ductile. Thus, as the composite is stressed, the load will be transferred from the matrix to the fibres by the shear stresses at the interfaces.

Experiments show that composites containing short fibres behave essentially the same as those with continuous fibres but are always less strong. In this respect the relative length l to the diameter d of the fibres is of importance. The ratio l/d is termed the *aspect ratio*. For example, it is found that a composite made from discontinuous lengths of tungsten wire embedded in copper has about 95% of the strength of the continuous-wire composite if wires having an aspect ratio of about 40 are used.

The strength of the composite is very dependent on the strength of the individual fibres. It has been calculated that the fibre will break owing to the plastic flow of the matrix material if

$$\frac{l}{d} \leqslant \frac{\sigma_f}{2\tau_m}$$

where σ_f is the ultimate tensile strength of the fibre and τ_m is the shear stress in the matrix. When this happens, l/d becomes the *critical aspect ratio*.

It may be seen that the manufacture of satisfactory high-strength composite materials involves many problems, several of which, it may be added, have already been overcome. The main problem has been that of producing large quantities of suitable high-strength fibre of the required aspect ratios. On a large scale, carbon fibres have so far been the most satisfactory in this respect. There are also problems of handling these fibres during the composite fabrication to avoid damage. The fibres have to be aligned and uniformly dispersed in the matrix and good bonding must be achieved. But, nevertheless, once all the problems have been solved, composite materials hold great promise for the future.

Chapter Four

VISCOSITY

4.1 Introduction

In this chapter we are concerned with the flow of liquids. We shall restrict ourselves to incompressible liquids and leave compressible fluids, gases, to the next chapter. Also to be left to the next chapter are those substances such as wet sand that can be labelled as fluids but which do not behave in any simple manner. This family of fluids can come under the broad heading of non-Newtonian fluids.

Viscosity is the resistance to flow of a fluid and may be thought of as internal friction. In Figure 4.1 we consider a liquid flowing over a horizontal surface. The layer adjacent to the horizontal surface we can imagine as being stationary, owing perhaps to friction between the

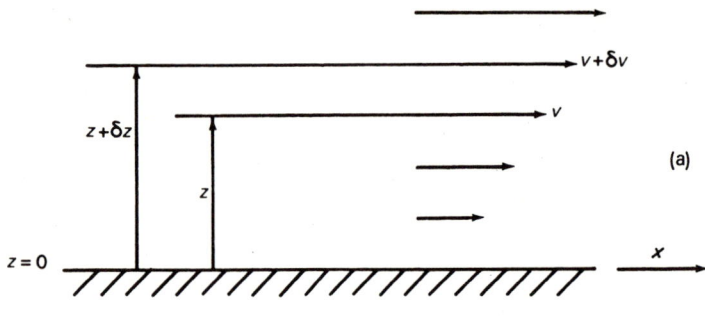

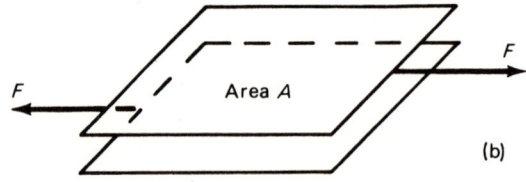

Figure 4.1. *Illustrating the flow of liquid over a surface*

liquid and the solid or owing to molecular attraction. A layer of liquid further up will, however, be moving, as will all the other layers into which the liquid may be imagined divided. Each layer will be slipping over the one beneath it with a velocity that is greater the further it is from the horizontal fixed surface. How easily each layer slips over the one beneath depends on the friction between the layers or, in other words, on the viscosity. This type of flow is termed *laminar* or, better still, *streamline flow*, since the opposite of turbulent flow is stream-lined but not necessarily laminar. A streamline or flow line is the path taken by a particle moving freely with the liquid. For the flow to be streamlined we mean that, at any point in the moving liquid, the pressure, and the magnitude and direction of the velocity of the liquid, will remain constant, so any particles put into the liquid at that point would always move along the same path. In other words, there is no turbulence producing unpredictable flow paths.

If two layers of liquid, distance z and $z + \delta z$ from the fixed horizontal surface, have velocities v and $v + \delta v$ respectively, as shown in Figure 4.1, then dv/dz is termed the *velocity gradient*. The upper layer also exerts a shearing stress on the layer beneath equal to F/A, where F is the shearing force acting over the common area A between the layers, as shown in Figure 4.1(b). It is, of course, this shearing stress that is causing the liquid to flow in the first place. According to Newton this shear stress is proportional to the velocity gradient it produces, providing the flow remains streamlined. That is,

$$\frac{F}{A} = \eta \frac{dv}{dz} \tag{4.1}$$

where η, the constant of proportionality, is termed the coefficient of viscosity, or simply the *viscosity*. It is also termed the 'dynamic' viscosity. From equation 4.1,

$$\eta = \frac{F}{A\,(dv/dz)}$$

that is, η has the mass, length, and time dimensions of

$$\frac{[MLT^{-2}]}{[L^2]\,[LT^{-1}/L]} = [ML^{-1}\,T^{-1}]$$

and thus units of kg m^{-1} s^{-1}. This is normally expressed in units of newton second per metre squared (N s m^{-2}). The reciprocal of η is termed the coefficient of fluidity. Another unit in common use is that

123

of the *kinematic viscosity v*, equal to η/ρ where ρ is the density of the liquid. This has the units metre squared per second ($m^2 s^{-1}$). The older unit for viscosity was the poise, equal to 1 g cm^{-1} s^{-1} (10^{-1} N s m^{-2}), and its hundredth part the centipoise, and for the kinematic viscosity, the stokes, equal to 1 cm^2 s^{-1} (10^{-4} m^2 s^{-1}). In Table 4.1 the viscosity is given for a number of common liquids.

Table 4.1. VISCOSITIES OF SOME COMMON LIQUIDS

Liquid	Dynamic viscosity, η (10^{-3} N s m^{-2} or centipoise)			Kinematic viscosity, v (10^{-6} m^2 s^{-1} or centistokes)		
	0°C	20°C	40°C	0°C	20°C	40°C
Acetone	0·40	0·32	0·27	0·50	0·41	0·34
Benzene	0·91	0·65	0·49	1·03	0·74	0·56
Carbon tetrachloride	1·35	0·97	0·74	0·85	0·61	0·46
Castor oil	6400	973	227	6640	1008	235
Chloroform	0·70	0·57	0·47	0·47	0·38	0·31
Ethyl alcohol	1·78	1·19	0·825	2·21	1·51	1·07
Mercury	1·68	1·55	1·45	0·12	0·11	0·11
Methyl alcohol	0·82	0·58	0·45	1·01	0·73	0·58
Trichlorethylene	0·71	0·58	0·48	0·49	0·40	0·33
Water	1·786	1·002	0·654	1·786	1·004	0·659

In this chapter we shall restrict the discussions to those liquids that obey Newton's law of viscosity (equation 4.1), that is, the so-called 'Newtonian' liquids.

4.2 Viscosity and Elasticity

In Chapter 1 we derived a number of equations that described and related the elastic properties of a solid. To describe the properties fully we had to make use of tensors. The same applies to liquids but, owing to the nature of a liquid, some simplification is possible and there are less independent terms. Let us then draw some analogies between a flowing liquid and an elastic solid.

If we consider the liquid flowing over the horizontal surface shown in Figure 4.1(a), we can see that the liquid is being sheared in much the same way that we considered a solid cube to be sheared when we defined the shear modulus of elasticity in Section 1.1. The only difference is that with the liquid the shearing continues at a constant rate with a constant applied shearing force. By definition the shear modulus G (equation 1.5) is given by

$$G = \frac{\text{shear stress}}{\text{shear strain}}$$

$$= \frac{\text{tangential force per unit area, } \tau}{\gamma} \tag{4.2}$$

as illustrated in Figure 4.2(a) where a cube of face area A is sheared through a small angle γ such that $\tan \gamma$ is approximately equal to γ and

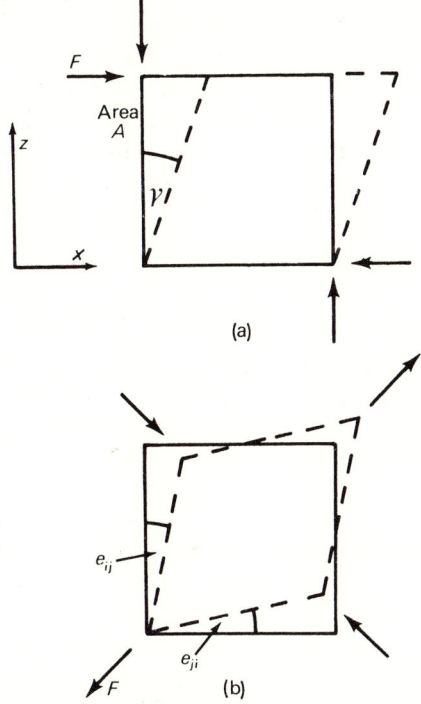

(a)

(b)

Figure 4.2.   Showing the relation between the engineering shear strain γ and the tensor shear strain e_{ij}

$\tau = F/A$. We also showed in equation 1.35 that the tensor shear strain is half the engineering shear strain occurring in the definition of the shear modulus of rigidity (equation 4.2). That is,

$$\epsilon_{ij} = \frac{1}{2} \gamma \tag{4.3}$$

125

ϵ_{ij} is the tensor shear strain and is illustrated in Figure 4.2(b) where the sides of the cube are rotated as it is sheared and

$$\epsilon_{ij} = \frac{1}{2}(e_{ij} + e_{ji}) \tag{4.4}$$

Thus, from equations 4.2 and 4.4,

$$2\,\epsilon_{ij} = \gamma = \frac{\tau}{G} \tag{4.5}$$

Now, if we consider a flowing liquid, a shear stress will produce a deformation that increases without limit. In other words, the deformation ϵ_{ij}, the tensor shear strain, produced in a solid may by analogy be replaced by a tensor flow, f_{ij}, in the case of a liquid. Also, for a solid, γ in Figure 4.2(a) is equal to dv/dz, whereas for a liquid which is flowing we are concerned with dv/dz as in Figure 4.1(a), equal to $d\gamma/dt = \dot{\gamma}$. Hence, by analogy with equation 4.5, we can write equation 4.1 as

$$2f_{ij} = \dot{\gamma} = \frac{\tau}{\eta} \tag{4.6}$$

where τ, the shear stress, is equal to F/A where the force F is in the plane of the area A, as in Figure 4.1(b). The analogy thus makes it possible to apply equations derived for an isotropic solid to a Newtonian liquid.

We can, for example, write the equation for Young's modulus E, that is,

$$E = \frac{\sigma}{\epsilon} \tag{4.7}$$

where σ is the longitudinal stress and ϵ the resulting longitudinal strain, analogously for a liquid as

$$\lambda = \frac{\sigma}{f_{ii}} \tag{4.8}$$

where λ is termed the Trouton coefficient of viscous traction and f_{ii} is the extensional flow in the direction of a longitudinal stress. This is, of course, mainly applicable to very viscous liquids that may be formed into a rod specimen and then stretched. For a solid we showed in Chapter 1 that the moduli of elasticity are connected by the expression (equation 1.55)

$$E = \frac{9KG}{3K + G} = 3G - \frac{G^2}{K} + \frac{G^3}{3K^2} \cdots \qquad (4.9)$$

where E is Young's modulus and K and G are the bulk and shear moduli respectively. For an incompressible material, that is, where $K = \infty$,

$$E = 3G \qquad (4.10)$$

from equation 4.9. By analogy then for a liquid, which it may be assumed is incompressible,

$$\lambda = 3\eta \qquad (4.11)$$

Thus we see from these few examples that there is a close relationship between the theory developed for elastic bodies and that applied to flowing bodies.

4.3 Flow through Tubes

Equations do exist, for example the Navier–Stokes equations, which describe the motion of a flowing liquid in all circumstances, but these equations are complex and in most cases are very difficult or even impossible to solve. Such equations have greater use in aerodynamics but for liquids we are generally concerned in practice with flow through a cylindrical tube or pipeline, which makes possible a great simplification of the mathematical treatment. The flow rate of a liquid may be related to the geometry of the tube by means of the Hagen–Poiseuille equation, which we shall now derive.

In discussing the flow of liquid through a tube we must restrict the case to only streamline flow, otherwise there would be turbulence. A theoretical treatment would then become impossibly complex as the flow direction and velocity of the liquid at any point would be continuously changing. For streamline flow we may assume that the layer of liquid in contact with the tube wall is stationary, in the same way as when we discussed flow over a horizontal surface in connection with Figure 4.1. Also there is no radial movement of liquid; it flows straight through the tube. As the liquid is stationary by the tube wall, the flow may be imagined as telescopic so that each concentric tube-like layer of liquid slips through relative to its neighbours. The 'tube' with the smallest radius at the centre of the system will move with the greatest velocity.

In Figure 4.3 is shown a tube of radius r and length l. So that gravity will not influence the flow rate, we shall consider the tube to be horizontal. Suppose now that there is a constant pressure gradient along the tube of $l/(p_1 - p_2)$, due to pressures p_1 and p_2 acting over vertical cross-sections in the liquid, distance l apart, and where $p_1 > p_2$. It is this pressure gradient that is causing the liquid to flow through the tube.

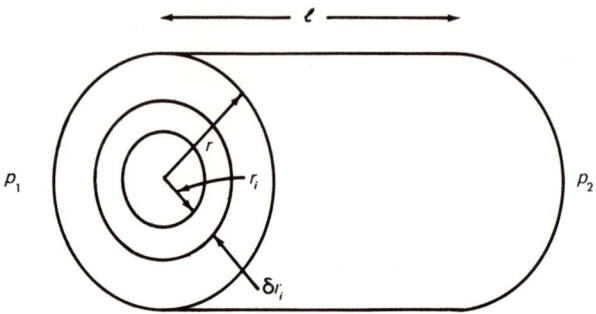

Figure 4.3.   Illustration for calculations on flow of a liquid through a tube

Let us now consider two elemental tubular layers of liquid concentric with the tube walls, as shown in Figure 4.3, such that one elemental tube has a radius r_i and moves with a velocity v_i, and the other a radius $r_i + \delta r_i$ and moves with a velocity $v_i + \delta v_i$. Thus there is a velocity gradient dv_i/dr_i between them and a consequent tangential stress $\eta\, dv_i/dr_i$, where η is the viscosity of the liquid. According to Newton's hypothesis (equation 4.1), there is a retarding viscous force of $\eta A\, dv_i/dr_i$ where A is the common area of contact between the elemental layers, that is,

$$\text{retarding viscous force} = 2\pi r_i\, l\, \eta \frac{dv_i}{dr_i} \qquad (4.12)$$

Now we have already said that it is the pressure difference $p_1 - p_2$ between points l apart that is causing the liquid to flow. Thus the force that causes the cylinder of liquid of radius r_i to slide through the tube of radius $r_i + \delta r_i$ is $\pi r_i^2 (p_1 - p_2)$, so that for steady flow

$$\pi r_i^2 (p_1 - p_2) = -2\pi r_i\, l\, \eta \frac{dv_i}{dr_i} \qquad (4.13)$$

where the negative sign indicates a retarding force. Therefore

$$\frac{dv_i}{dr_i} = -\frac{(p_1 - p_2)\, r_i}{2\, \eta\, l} \qquad (4.14)$$

128

and

$$\int dv_i = -\frac{p_1 - p_2}{2\eta l} \int r_i \, dr_i \qquad (4.15)$$

Therefore

$$v_i = -\frac{p_1 - p_2}{2\eta l} \left(\frac{r_i^2}{2}\right) + \text{const.} \qquad (4.16)$$

However, at the wall of the tube we know that the liquid is stationary, that is, $v_i = 0$ at $r_i = r$, so the constant of equation 4.16 is

$$\text{const.} = +\frac{p_1 - p_2}{2\eta l} \left(\frac{r^2}{2}\right) \qquad (4.17)$$

and therefore, from equations 4.16 and 4.17,

$$v_i = \frac{p_1 - p_2}{4\eta l} (r^2 - r_i^2) \qquad (4.18)$$

where v_i is the velocity of the liquid at a radius r_i. It can also be seen from this equation that the velocity profile across the tube is a parabola, as shown in Figure 4.4.

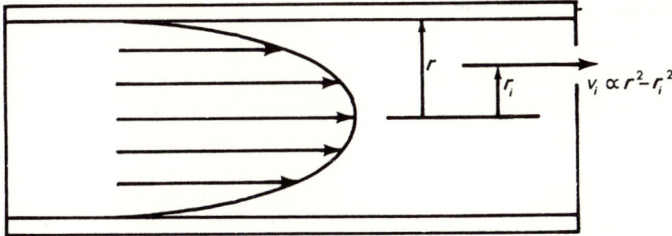

Figure 4.4. *Parabolic velocity profile in the streamlined flow of a liquid through a tube*

The velocity is not a very practical quantity to determine so we must obtain a more useful expression — that for the *volume* of liquid passing through the tube per second. We know that the volume of liquid δQ flowing along the tube per second through the annulus bounded by the layers of radii r_i and $r_i + \delta r_i$ is

$$\delta Q = 2\pi r_i v_i \delta r_i \qquad (4.19)$$

where $2\pi r_i \delta r_i$ is the area of the annulus. Hence the total volume of liquid flowing through the tube per second is

$$Q = \int_0^r \frac{2\pi r_i \, (p_1 - p_2) \, (r^2 - r_i{}^2)}{4 \eta l} \, dr_i$$

$$= \frac{\pi(p_1 - p_2)}{2\eta l} \int_0^r (r^2 r_i - r_i{}^3) \, dr_i$$

$$= \frac{\pi(p_1 - p_2)}{2\eta l} \left[\frac{r^2 r_i{}^2}{2} - \frac{r_i{}^4}{4} \right]_0^r$$

Therefore

$$Q = \frac{\pi(p_1 - p_2) \, r^4}{8\eta l} \qquad (4.20)$$

It should be noted that equation 4.20 can also be written in the form

$$Q = \frac{N \, (p_1 - p_2) \, r^4}{\eta l} \qquad (4.21)$$

where N is a numerical factor that will depend on the shape of the tube cross-section, which is not restricted to any particular geometric shape; r is then a function of the size of the cross-section. Except for including the viscosity in the factor N, this is the form of the equation first derived experimentally by Hagen and later by Poiseuille. Hagenbach then showed by theoretical reasoning that $N = \pi/8$ for a tube of circular cross-section and radius r.

From equation 4.18 the maximum velocity of liquid flow is along the axis of the tube, that is, when $r_i = 0$, so the maximum velocity is given by

$$v_{\max} = \frac{(p_1 - p_2) \, r^2}{4\eta l}$$

and therefore equation 4.20 can be written as

$$Q = \pi r^2 \left(\frac{v_{\max}}{2} \right) \qquad (4.22)$$

If we define the average velocity of flow $\bar{v}$ as the total volume of liquid flowing in unit time divided by the cross-sectional area of the tube, then

$$\bar{v} = \frac{Q}{\pi r^2} = \frac{v_{\max}}{2}$$

We may then take this as a characteristic of streamline flow through a tube — that the average velocity is half the maximum velocity.

130

Thus, if the viscosity is known, the rate of flow may be determined, or conversely equation 4.20 may be used to calculate the viscosity from measurements of the flow rate. Figure 4.5 shows a typical laboratory

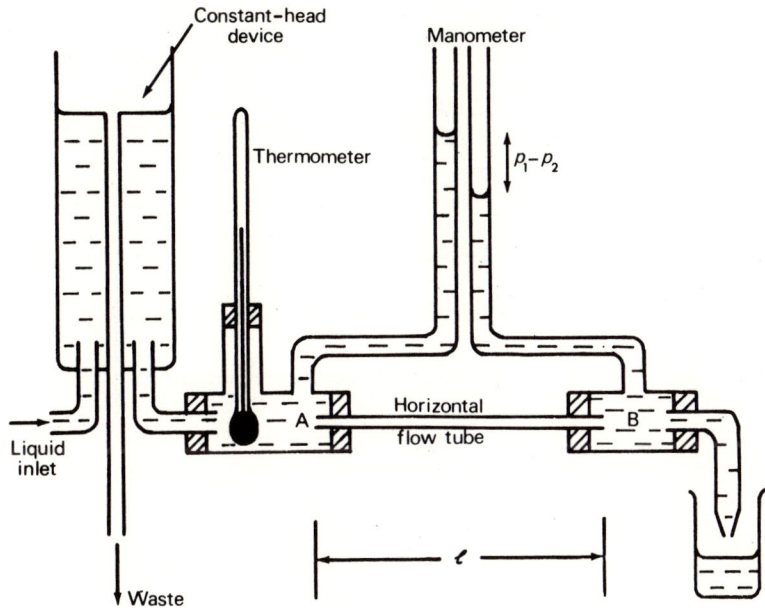

Figure 4.5.   *Apparatus for the measurement of viscosity by the Hagen–Poiseuille method*

arrangement for determining the parameters of equation 4.20. The volume flowing is best determined by weighing, the density being known, and the mean radius of the flow tube by measurements of a mercury thread.

Providing only small flow rates are involved, equation 4.20 is very nearly exact. However, two important factors have been neglected and for more precise work these must be taken into account. These corrections were suggested by Hagenbach. The first correction arises from the fact that accelerations occur in the region of the inlet to the tube, causing some radial movement of the liquid until it is far enough into the tube to have settled down into streamlined flow with flow lines parallel to the axis of the tube. To correct for this error an amount nr must be added to the measured length l of the tube, where r is the radius of the tube and n is an experimentally determined constant; n is equal to approximately 1·64 when the flow tube butts straight into the liquid reservoir, as in Figure 4.6, and when the pressure difference is taken simply as that due to the head of liquid. For the apparatus of Figure 4.5, n will tend to zero as the pressure chambers A and B tend to the same radius as the flow tube.

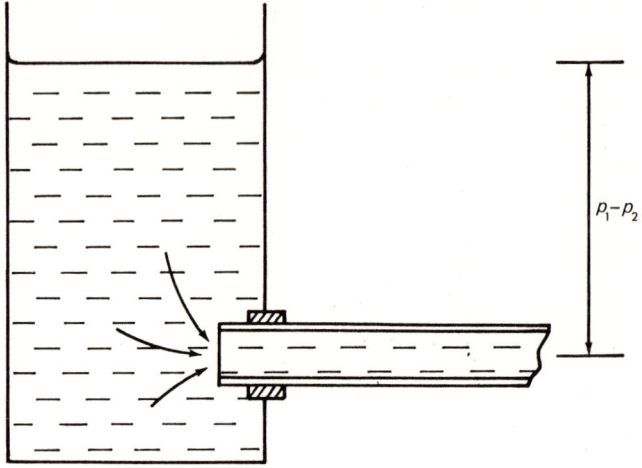

Figure 4.6. *Experimental arrangement where the flow tube butts straight into the reservoir*

The other necessary correction arises from the fact that we assumed that the pressure differential $p_1 - p_2$ was used solely in providing the force to overcome the viscous drag. In fact, of course, it also gives kinetic energy to the flowing liquid. Again this error will tend to zero if the pressure chambers A and B in the arrangement shown in Figure 4.5 tend to the same radius as the flow tube, since the manometer will simply record the pressure differential of the already flowing liquid. If an experimental arrangement as suggested in Figure 4.6 is used, a corrected expression can be derived for the pressure differential to take account of the kinetic energy error, as follows.

Referring back to Figure 4.3, we see that the volume of liquid flowing per second across the annulus of radius r_i and $r_i + \delta r_i$ is (equation 4.19) $\delta Q = 2\pi r_i v_i \delta r_i$. If the liquid has a density ρ, then the kinetic energy $(\frac{1}{2} m v_i^2)$ of the liquid flowing across this annulus is

$$\text{kinetic energy} = \frac{1}{2} (2\pi \rho \, r_i v_i \, \delta r_i) \, v_i^2$$

$$= \pi \rho \, v_i^3 r_i \, \delta r_i \qquad (4.23)$$

132

where v_i is given by equation 4.18, that is,

$$v_i = \frac{p_1{}' - p_2{}'}{4\eta l}(r^2 - r_i^2) \tag{4.24}$$

$p_1{}' - p_2{}'$ is the pressure differential required to overcome the viscous drag only and is less than the measured pressure differential $p_1 - p_2$ which has to provide the kinetic energy as well. The total kinetic energy for all the liquid flowing through the tube in one second, taken over a complete cross-section, is

$$\int_0^r \pi\rho\, v_i^3 r_i\, dr_i = \pi\rho \left(\frac{p_1{}' - p_2{}'}{4\eta l}\right)^3 \int_0^r (r^2 - r_i^2)^3\, r_i\, dr_i$$

$$= \pi\rho \left(\frac{p_1{}' - p_2{}'}{4\eta l}\right)^3 \left[\frac{r^6 r_i^2}{2} - \frac{3r^4 r_i^4}{4} + \frac{3r^2 r_i^6}{6} - \frac{r_i^8}{8}\right]_0^r$$

$$= \pi\rho \left(\frac{p_1{}' - p_2{}'}{4\eta l}\right)^3 \frac{r^8}{8} \tag{4.25}$$

However, since $p_1{}' - p_2{}'$ is the pressure differential required to overcome the viscous drag,

$$\frac{\pi(p_1{}' - p_2{}')\, r^4}{8\eta l} = Q \tag{4.26}$$

in agreement with equation 4.20. Therefore, substituting in equation 4.25,

$$\text{kinetic energy} = \frac{\rho Q^3}{\pi^2 r^4} \tag{4.27}$$

The total work expended per second must equal the work done in overcoming the viscous drag, plus the kinetic energy retained by the liquid as it leaves the flow tube. The work done in overcoming the viscous drag is equal to the pressure $p_1{}' - p_2{}'$ times the volume of liquid flowing per second, Q. Therefore

$$(p_1 - p_2)\, Q = (p_1{}' - p_2{}')\, Q + \frac{\rho Q^3}{\pi^2 r^4}$$

That is,

$$p_1{}' - p_2{}' = (p_1 - p_2) - \frac{\rho Q^2}{\pi^2 r^4} \tag{4.28}$$

where $p_1 - p_2$ is the measured pressure differential and $p_1{}' - p_2{}'$ is the effective pressure differential used in overcoming the viscous drag.

Hence equation 4.20 can be written, taking into account the corrections, and adding an extra multiplying constant m to take account of any small factors neglected in the theory, as

$$Q = \frac{\pi r^4}{8\,\eta\,(l + 1 \cdot 64r)} \left[(p_1 - p_2) - \frac{m\,\rho\,Q^2}{\pi^2 r^4} \right]$$

$$= \frac{\pi^2 r^4 (p_1 - p_2) - m\rho\,Q^2}{8\,\pi\eta\,(l + 1 \cdot 64r)} \tag{4.29}$$

That is,

$$\eta = \frac{\pi^2 r^4 (p_1 - p_2) - m\rho\,Q^2}{8\,\pi Q\,(l + 1 \cdot 64r)} \tag{4.30}$$

In practice m is approximately unity but should be determined by calibration for a particular experimental arrangement used in any very precise work.

4.4 Ostwald's Viscometer

Although the Hagen–Poiseuille method enables accurate determinations of viscosity to be made, it does have the disadvantages that it is slow, involves a considerable volume of liquid, and the apparatus and liquid are difficult to maintain at any specified temperatures that may be required. In addition it is found in practice that it is often easier to obtain accurate values of viscosity by relative measurements than by absolute determinations. A method involving relative measurements makes use of Ostwald's viscometer, commonly used in routine determinations. It is shown diagrammatically in Figure 4.7. Normally made of glass, it can vary in length and bore, depending on the range of viscosities to be measured, but on average is about 20 cm long. It is used vertically and has lines etched on at A, B, and C in the figure.

   A standard volume of liquid is pipetted into the right-hand tube and then blown or sucked up the left-hand tube. The volume of liquid is such as to occupy the space between the marks at A and C. With the level of liquid starting just above the mark A, the time t is found for the liquid level to fall vertically under gravity from A to B. The method really involves the Hagen–Poiseuille equation (equation 4.20), but takes into account the varying pressure differential. As the flow rate Q is equal to the volume V of liquid passing through the tube in time t, that is, $Q = V/t$, then equation 4.30 can be written as

$$\eta = \frac{\pi r^4 (p_1 - p_2)t}{8V(l + 1.64r)} - \frac{m\rho V}{8\pi(l + 1.64r)t}$$

$$= A\rho t - \frac{B\rho}{t} \qquad\qquad (4.31)$$

where A and B are constants depending on the particular viscometer. The pressure differential $p_1 - p_2$ has been written in the form $g\rho h$. By using two liquids of known viscosities, these constants A and B may be

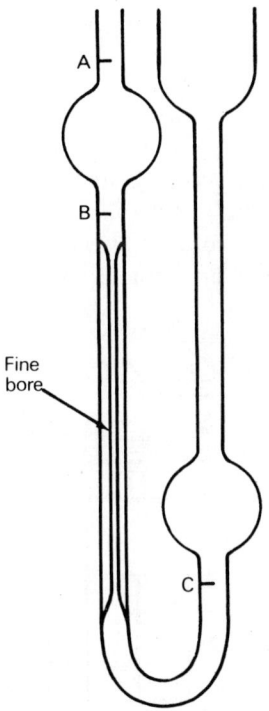

Fine bore

Figure 4.7. An Ostwald viscometer

determined although in practice the viscometers are commonly designed to make the correction term $B\rho/t$ negligible by ensuring a sufficiently long flow time for its particular viscosity range. In this case, of course, only one calibrating liquid is required – to determine A. The viscometer having been calibrated, the viscosity of an unknown liquid can be

135

simply determined by timing its fall from level A to level B and substituting this time in equation 4.31.

This method is obviously quick and convenient and enables repeated measurements to be made on a comparatively small amount of liquid. It also has the great advantage that it can conveniently be mounted in a thermostatted temperature bath so that measurements may be made on the same sample of liquid at various temperatures. The viscometer may also be used for determining the viscosity of molten metals by making it of a suitable material such as quartz or platinum and determining the levels by electrical contact methods.

4.5 Rotating Cylinder Viscometer

A different approach to the measurement of viscosity is to measure the torque required to rotate a vertical cylinder with a constant angular velocity in a liquid and thus to determine the rate of shear. This method is most suitable for very viscous liquids where the time taken to flow through a tube would prohibit methods involving the Hagen–Poiseuille equation.

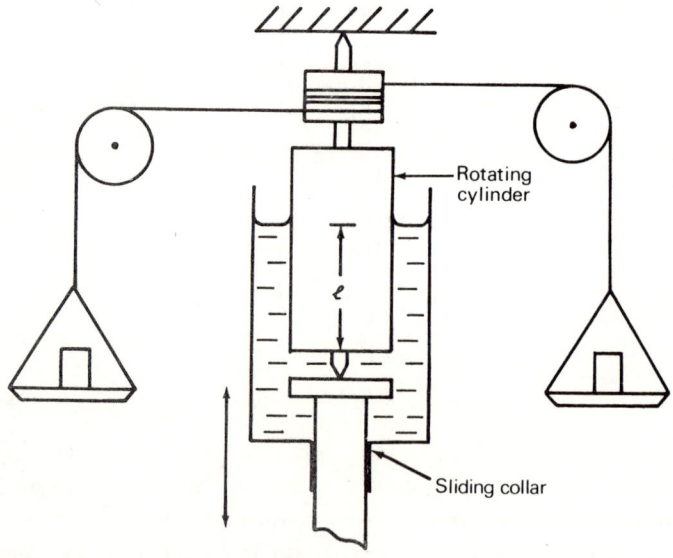

Figure 4.8.   Arrangement of a rotating cylinder viscometer

A common form of rotating cylinder viscometer consists of an inner cylinder of known diameter able to rotate within a fixed outer concentric cylinder. The annular space between is filled with the liquid under test to a known height up the inner cylinder, which is rotated at a constant velocity by a steadily applied couple due to an arrangement such as that shown diagrammatically in Figure 4.8. The height of liquid up the inner cylinder can be varied by sliding the outer cylinder up or down on its central support pillar.

We can derive the theory of the method in the following way. Figure 4.9 represents a cross-section of the cylinders and the annulus of liquid, where the inner rotating cylinder and the outer stationary

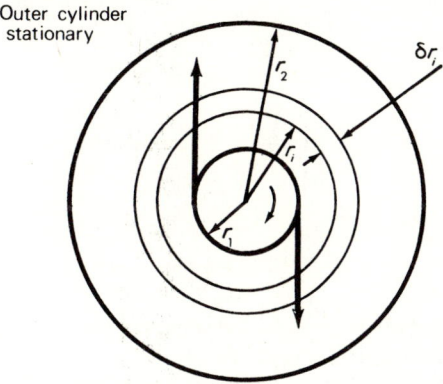

Outer cylinder
stationary

Figure 4.9.   Cross-section of a rotating cylinder viscometer

cylinder are of radii r_1 and r_2 respectively. Let us consider an elemental annular tube of liquid of radius r_i and thickness δr_i. If the inner cylinder is rotated at a constant angular velocity, an angular velocity ω will be given to the layer of liquid of radius r_i. That is, the liquid will have a linear velocity ωr_i. There will thus be a velocity gradient across the elemental annulus of

$$\frac{d}{dr_i}(\omega r_i) = r_i \frac{d\omega}{dr_i} + \omega \qquad (4.32)$$

The last term in this equation is just the angular velocity of the layer, and it is the term $r_i \, d\omega/dr_i$ that gives rise to the viscous stress. Then, from Newton's equation (equation 4.1), the shearing force between the layers r_i and $r_i + \delta r_i$ is

$$F = -2\pi r_i \, l \eta \left(r_i \frac{d\omega}{dr_i} \right) \qquad (4.33)$$

137

where $2\pi r_i l$ is the common area between the layers and l is the length of the cylinders. The negative sign arises since the angular velocity ω decreases as the radius increases. Thus the torque required to rotate the layer of radius r_i relative to the layer of radius $r_i + \delta r_i$ is

$$T = Fr_i = -2\pi\eta l\, r_i^3 \frac{d\omega}{dr_i} \tag{4.34}$$

In the equilibrium state, this is the torque applied to the inner cylinder. Since the liquid in contact with the outer cylinder may be regarded as stationary, that is, $\omega = 0$ at $r_i = r_2$, and taking the angular velocity to be ω at $r_i = r_1$, we may integrate between these limits. That is,

$$-T\int_{r_1}^{r_2} \frac{1}{r_i^3}\, dr_i = 2\pi\eta l \int_{\omega}^{0} d\omega \tag{4.35}$$

or

$$T\left[\frac{1}{2r_i^2}\right]_{r_1}^{r_2} = 2\pi\eta l\, [\omega]_{\omega}^{0}$$

and therefore

$$T = 4\pi\eta l\omega\left(\frac{r_1^2 r_2^2}{r_2^2 - r_1^2}\right) \tag{4.36}$$

where T is the torque applied to the inner cylinder. This is the Margules equation, which if written in the form

$$\omega = \frac{T}{4\pi\eta l}\left(\frac{1}{r_1^2} - \frac{1}{r_2^2}\right)$$

can be compared and constrasted with equation 4.18 for the velocity of a liquid flowing through a tube.

We have, however, neglected in equation 4.36 the drag due to the horizontal layer of liquid under the base of the inner cylinder, but can eliminate its effect by considering two lengths of immersion l_1 and l_2 of the inner cylinder in the liquid. If the rotation velocity is the same in both cases, then the torque required to overcome this end-effect drag will also be the same. Let this extra torque be T', then

$$T_1 = 4\pi\eta l_1\omega\left(\frac{r_1^2 r_2^2}{r_2^2 - r_1^2}\right) + T' \tag{4.37}$$

and

$$T_2 = 4\pi\eta l_2\omega\left(\frac{r_1^2 r_2^2}{r_2^2 - r_1^2}\right) + T' \tag{4.38}$$

138

where T_1 and T_2 are the torques required in the two cases to produce the same angular velocity ω. Subtracting one equation from the other,

$$T_1 - T_2 = \frac{4\pi\eta\,\omega\,r_1{}^2 r_2{}^2\,(l_1 - l_2)}{r_2{}^2 - r_1{}^2} \tag{4.39}$$

As in Figure 4.8, the two immersion lengths l_1 and l_2 are easily achieved by sliding the outer cylinder vertically on its supporting pillar. In practice, very low angular velocities should be avoided as the friction of the bearings then becomes an important consideration.

4.6 Stokes' Falling-sphere Viscometer

This technique involves the rate at which a spherical ball falls under gravity through a liquid or, for that matter, a gas. The exact theoretical derivation is of great mathematical difficulty, so we shall make use of a simpler approach, that of dimensional analysis, although this cannot give the value of the numerical constant occurring in the final equation. The problem is as follows. A small sphere is falling through a viscous medium. At first its velocity will increase but, when the force on the sphere due to gravity becomes balanced by the viscous drag and the buoyancy of the sphere, the velocity will no longer increase and will remain constant. The falling body has then reached its *terminal velocity*.

It is reasonable to assume that the viscous retarding force F is some function of the size of the falling sphere of radius r, its velocity v, and of course the viscosity of the medium η. That is,

$$F = Nr^\alpha\,v^\beta\,\eta^\gamma \tag{4.40}$$

where N is a numerical constant of proportionality. The indices α, β, and γ can be determined by dimensional analysis. Expressing equation 4.40 in terms of its dimensions,

$$[M\,L\,T^{-2}] = [L]^\alpha\,[L\,T^{-1}]^\beta\,[M\,L^{-1}\,T^{-1}]^\gamma$$

that is,

$$[M]^1\,[L]^1\,[T]^{-2} = [M]^\gamma\,[L]^{\alpha+\beta-\gamma}\,[T]^{-\beta-\gamma}$$

139

Equating indices,

$$1 = \gamma$$

$$1 = \alpha + \beta - \gamma$$

$$-2 = -\beta - \gamma$$

from which

$$\alpha = \beta = \gamma = 1$$

and therefore equation 4.40 becomes

$$F = Nrv\eta \tag{4.41}$$

Stokes showed by a full theoretical treatment that the numerical factor $N = 6\pi$, and

$$F = 6\pi rv\eta \tag{4.42}$$

When a falling sphere has reached its terminal velocity v, then the viscous retarding force and the buoyant force must be balanced by the gravitational force on the sphere, that is,

$$6\pi rv\eta + \frac{4}{3} \pi r^3 \rho_L g = \frac{4}{3} \pi r^3 \rho_S g$$

or

$$\eta = \frac{2\, r^2 g}{9v} (\rho_S - \rho_L) \tag{4.43}$$

where ρ_S and ρ_L are the densities of the sphere and liquid respectively, r is the radius of the sphere, and v is its terminal velocity under an intensity of gravity g.

To determine the viscosity of a liquid experimentally by an application of equation 4.43, care must be taken to ensure that the sphere has truly reached its terminal velocity and that side and end effects are minimised. In the full theoretical derivation, Stokes assumed that the sphere was falling through an infinite ocean of liquid. For a finite volume of liquid, the walls of the containing vessel will have an effect. For a cylindrical vessel, the falling sphere will displace liquid that has to flow through the annulus between the sphere and the walls. Similarly, disturbing effects occur near to the surface of the liquid and the bottom of the vessel. Ladenburg has shown that the true terminal velocity v_∞ occurring with an infinite ocean of liquid is related to the observed velocity v

140

associated with a finite volume by

$$v_\infty = v\left(1 + 2.4\frac{r}{R}\right) \qquad (4.44)$$

where r is the radius of the sphere and R is the radius of the cross-section of the cylindrical vessel containing the liquid. This corrects for the side effect providing the sphere falls along the axis of the cylinder. To take account of the end effects, Ladenburg has suggested that the height of the column of liquid should be divided into three parts of which the central third should be used for the measurements of the terminal velocity. Then as before the true terminal velocity v_∞ is related to the observed terminal velocity v by

$$v_\infty = v\left(1 + 3.3\frac{r}{h}\right) \qquad (4.45)$$

where h is the total height of the column of liquid. Generally this correction is very small. Combining these corrections with equation 4.43,

$$\eta = \frac{2r^2g\,(\rho_S - \rho_L)}{9v\,[1 + 2.4\,(r/R)]\,[1 + 3.3\,(r/h)]} \qquad (4.46)$$

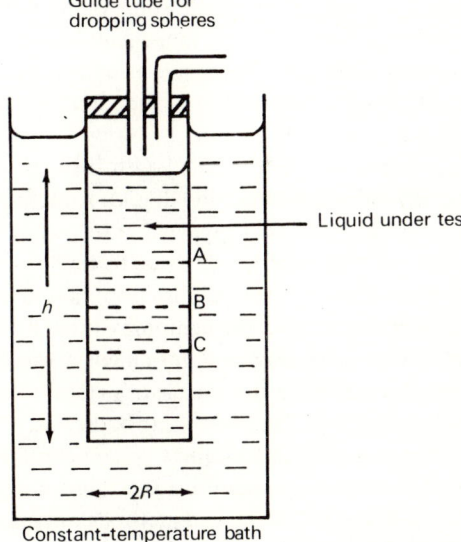

Figure 4.10. Apparatus for the determination of viscosity by Stokes' method

Providing that R/r is greater than about 200, and neglecting end effects as very small, the uncorrected Stokes' equation (equation 4.43) gives the viscosity correct to about one per cent. Large errors, however, can result owing to departures from sphericity of the falling balls and consequently such departures should be avoided.

A typical experimental arrangement for determining the viscosity by Stokes' method is shown in Figure 4.10. The spheres are dropped through a guide tube to ensure that they fall along the axis of the cylinder. The velocity is determined by timing the spheres between A and C. By measuring the times taken between A and B and between B and C, where B is midway between A and C, a check may be made to ensure that the spheres have previously reached their terminal velocity. Since the viscosity is dependent on temperature, a constant-temperature bath is required to obtain specific temperatures.

4.7 Other Methods for Determining Viscosity of Liquids

There have been many methods devised for determining the viscosity of liquids by an application of Newton's basic equation (equation 4.1); many of these, however, have tended to be academic exercises rather than serious practical methods. Some methods have involved departures from the Hagen—Poiseuille arrangement by, for example, using non-horizontal flow tubes. There have also been a number of variations in design of rotation viscometers — for example, one method involves the rotation of a thin disc.

Other methods are concerned with the damping of the vibration of some body when immersed in a liquid. A method due to Maxwell utilises a horizontal disc supported at its centre by a torsion wire. The period of oscillation of the disc due to torsion in the suspension is measured in air and when the disc is immersed in the liquid. From the change in period due to the damping caused by the viscous drag of the liquid, the viscosity may be determined. A variation on this is due to Stokes, who measured the change in period of a pendulum when immersed in liquid. Other vibrating bodies used have been a solid sphere in the liquid, and a hollow sphere containing liquid. Again the methods have been to determine the changes in period due to damping.

The oil industry has its own methods of measuring viscosity and also its own set of empirical units, which form the subject of the next section.

4.8 Viscosity of Oil

The oil industry has traditionally been concerned with the kinematic viscosity, that is, the ratio of the dynamic viscosity to the density at the same temperature. It has been common to express this in centistokes, one hundredth of a stokes, where one centistokes is 10^{-6} m^{2}s^{-1}. One of the major uses of lubricating oils is in motor vehicles, and therefore oils have been classified into groups of suitable viscosities for this usage. Of particular interest here is the viscosity of the oil when the vehicle is standing in the winter as this affects the starting characteristics. For this purpose the oil is tested at -18°C (0°F) and at 99°C (210°F), which is approximately the maximum temperature reached by the oil sump of most car engines. The Society of Automotive Engineers (SAE) has classified the viscosities of oils measured at these standard temperatures into SAE viscosity numbers. Some SAE numbers have the suffix W added to signify that the viscosity limits are specified at -18°C, whilst others have two numbers, such as 10W/30, which gives the classification at both the low and the high temperature. Table 4.2 gives the kinematic viscosity ranges in units of 10^{-6} m^2 s^{-1} for various SAE viscosity

Table 4.2. KINEMATIC VISCOSITY RANGES FOR OILS COMMONLY USED IN CAR ENGINES

SAE viscosity number	Kinematic viscosity range (10^{-6} m^2 s^{-1})			
	-18°C		99°C	
	min.	max.	min.	max.
10W	1 303	2 606	–	–
20W	2 606	10 423	–	–
20	–	–	5·73	9·62
30	–	–	9·62	12·93
40	–	–	12·93	16·77
50	–	–	16·77	22·68

numbers for oils commonly used in car engines. The viscosity of fuel oils is also of importance. For example, the viscosity of paraffin used in wick-fed heaters and lamps should not be too low because of the resulting high feed rate and the possibility of an increase in the flame height after warming up. Conversely, too high a viscosity restricts the flow, leading to charring of the wick. The viscosity of diesel fuel oil is likewise important because it affects the spray pattern from the injector nozzles and may lead to less efficient combustion.

It is thus seen that quick methods for routine measurement of viscosity are required in the oil industry. The usual laboratory method for precise work is to use an Ostwald-type viscometer. This has the advantage that the temperature of the viscometer and the oil under test is easily maintained at the required values by means of a constant-temperature oil bath. From equation 4.31, the kinematic viscosity $\eta/\rho = At$ where the correction term is regarded as negligible, hence the measurement merely requires the timing of the fall of oil level between two marks and then multiplication by the viscometer constant. By having a choice of viscometers available with different capillary bores, the measuring time required for any grade of oil can always be of a convenient period.

Also commonly used are the Redwood, Saybolt, and Engler visco-meters, which are all similar in principle but less accurate than the Ostwald type. The *Redwood viscometer* is essentially an oil cup fitted with an agate jet in the base. Removing the plug in the jet allows oil to flow, and the time is measured for 50 ml to be collected in a cali-brated flask. The viscosity is then given in Redwood seconds. Tempera-ture control is obtained by surrounding the oil cup with a water bath. Two sizes of this viscometer are used − one is intended to give 10 times the flow rate of the other. The *Saybolt viscometer* is similar to the Redwood in that an oil cup fitted with a jet is surrounded by a heating bath, but now the time is measured for 60 ml to flow through the jet into a calibrated flask. The viscosity is then given in Saybolt seconds. Again two versions are in use, one to give 10 times the flow rate of the other. Yet another is the *Engler viscometer*, working on the same principle as the Redwood and Saybolt viscometers but giving the viscosity in Engler degrees. The Engler viscometer is used mainly in continental Europe, whereas the Redwood and Saybolt viscometers are commonly used in Great Britain and America respectively.

There is no standard factor for converting these various units from one system to another, but a graphical conversion is given in Figure 4.11. However, the relationships between Redwood and Saybolt seconds and kinematic viscosity vary with temperature. Consequently conversions between these units as given in Figure 4.11 must be regarded as approximate. Accurate conversions can be made only by using tables that give values at standard temperatures. Unfortunately there appears to have been no agreement over the choice of standard temperatures between the systems but, using the more accurate published tables rather than the graph of Figure 4.11, interpolation can be made between

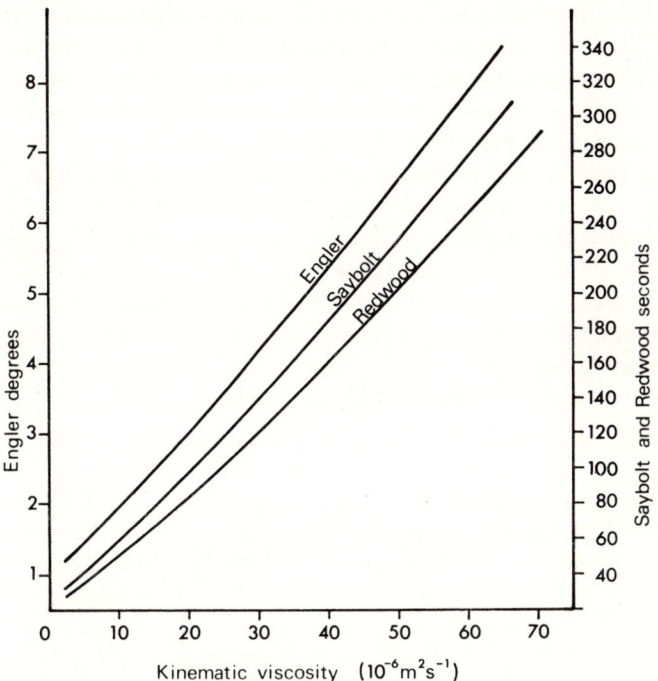

Figure 4.11. Approximate conversions of kinematic viscosity to Engler degrees, and Saybolt and Redwood seconds

temperatures. Conversion to Engler degrees from $m^2\ s^{-1}$ units involves negligible temperature variation. For conversions of viscosities above the range of those given in the graph, constant multiplying factors may be used but are not given here as they are of little interest to the general reader.

4.9 Variation of Viscosity of Liquids with Pressure and Temperature

Comparatively little work has been done in studying the effect of pressure on viscosity but it is known that the viscosity of all liquids, with the exception of water, increases rapidly with pressure. The relationship is approximately of the form

$$\eta = n \log_e p \qquad (4.47)$$

where p is the pressure and n a constant depending on the particular

145

liquid. Water exhibits an anomalous behaviour in that at temperatures between 0°C and about 10°C the viscosity falls to a minimum at a pressure of approximately 10^8 N m^{-2}. At higher temperatures there appears to be no minimum viscosity; the viscosity then increases with pressure in conformity with other liquids.

The viscosity of liquids falls rapidly with a rise in temperature. Consequently, in quoting values for viscosity it is essential that the temperature also is quoted. For example, water at 10°C has a viscosity of 13.09×10^{-8} N s m^{-2} dropping to 3.17×10^{-8} N s m^{-2} at 90°C. No complete theory has yet been derived to explain the variation with temperature for all liquids, but an expression due to Andrade is in good experimental agreement with the behaviour of most liquids except water and some of the alcohols. This expression is of the form

$$\eta = A \, \rho^{1/3} \exp\left(\frac{C\rho}{T}\right) \tag{4.48}$$

where ρ is the density of the liquid at the absolute temperature T, and A and C are constants for the particular liquid.

4.10 Turbulence

If the flow rate of a liquid is increased above a certain critical value, a change in the nature of the flow occurs. This would occur if, for example, the differential pressure or the diameter of the flow tube was increased in a Hagen–Poiseuille type experiment. Above a certain critical value of the velocity, termed the *critical velocity,* the layer of liquid in contact with the walls of the tube will still be stationary. The velocity will increase uniformly away from the wall for a short distance. Within the uniformly increasing velocity layer, termed the *boundary layer,* the flow is still streamlined. Beyond the boundary layer, the flow becomes *turbulent*. The motion is very irregular and random circular currents occur in localised regions. These random circular currents are called *vortices* and cause a large increase in the resistance to flow of the liquid. With turbulent flow the velocity profile changes from the parabolic form of Figure 4.4 to that shown in Figure 4.12.

It has been found by experiment that four parameters are relevant in deciding whether a flow shall be streamlined or turbulent. These parameters are combined to give the *Reynolds' number R*, such that

$$R = \frac{\rho \, vd}{\eta} \tag{4.49}$$

where ρ is the density of the liquid, v its mean velocity of flow, d the diameter of the tube, and η the viscosity of the liquid. It may be seen that this is a dimensionless number and has the same value irrespective of the system of units that may be used. For a Reynolds' number less

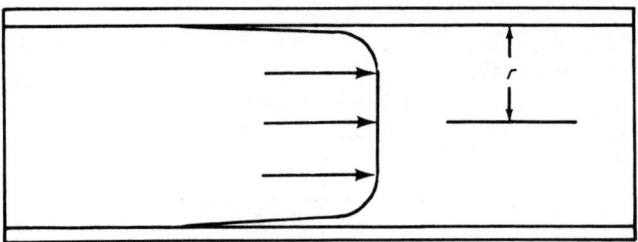

Figure 4.12.   Velocity profile in turbulent flow of a liquid through a tube

than about 2000 the flow is found to be streamlined, whereas for a value greater than about 3000 the flow is quite turbulent. Between 2000 and 3000 an unstable flow may be expected, changing from one type to the other.

Turbulence occurs with surprisingly slow flow rates. For example, if we take the onset of turbulence for water to occur with a Reynolds' number of about 2000, then from equation 4.49, substituting the appropriate values for the density and viscosity of water at, say, 20°C and assuming a tube diameter of 3 mm,

$$v = \frac{R\eta}{\rho d}$$

$$= \frac{2000 \times 1\cdot002 \times 10^{-3}}{1000 \times 0\cdot003}$$

$$= 0\cdot67 \text{ m s}^{-1}$$

Referring back to the Hagen—Poiseuille equation (equation 4.20) and assuming a tube length of, say, 0·25 m, we can calculate the pressure differential to produce this flow rate. That is,

$$p_1 - p_2 = \frac{8\eta l Q}{\pi r^4}$$

$$= \frac{8\eta l v}{r^2}$$

147

since $Q = \pi r^2 v$. Therefore

$$p_1 - p_2 = \frac{8 \times 1.002 \times 10^{-3} \times 0.25 \times 0.67}{0.0015^2}$$

$$= 597 \, N \, m^{-2}$$

$$= \rho gh$$

That is, a water head of

$$h = \frac{597}{\rho g}$$

$$= \frac{597}{1000 \times 9.8}$$

$$= 0.061 \, m$$

In other words, when measuring the viscosity of water by a Hagen–Poiseuille type experiment using a flow tube 0·25 m long with a 3 mm bore, it is necessary to keep the head of water to less than about 6 cm otherwise the flow may be turbulent and equation 4.20 would not be applicable.

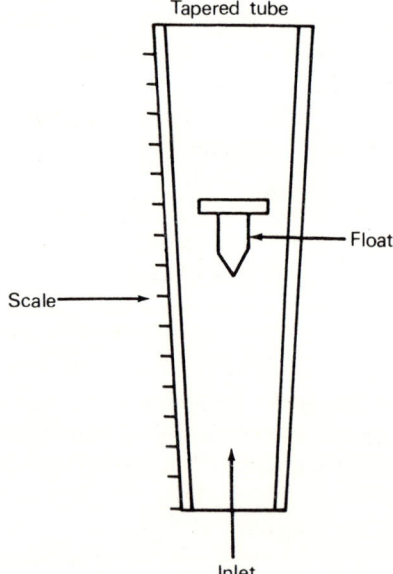

Figure 4.13. Showing the arrangement of a rotameter for measuring rates of flow of liquids and gases

As the flow rate is increased beyond the critical value, it does not continue to increase at the same rate owing to the increasing magnitude of the turbulence. At higher flow rates the flow becomes less dependent on viscosity and more dependent on density. This dependency on density of the flow rate for very turbulent flow is utilised in *rotameters.* These are tapered glass tubes, usually with a one-in-ten taper, in which can move a float, as shown in Figure 4.13. The shape of the float is such as to induce turbulence of the gas or liquid flowing through the vertical tube. As the flow rate is increased, the float is lifted higher up the tube, increasing the area of the annular gap between the float and the wall, so the vertical position taken up by the float is dependent on the flow rate through the tube. The tube carries a scale giving the flow rate from the position of the top of the float. Since the flow is very turbulent, the calibration of the scale is mainly dependent on the density of the gas or liquid. This fact extends the use of the rotameter by making it usable with a range of substances since, for example, many gases have very similar densities.

At even greater flow rates the dependency on the density changes to a dependency on the square root of the pressure, which is used to impart kinetic energy to the moving liquid. The fact that viscosity plays virtually no part in the flow rate of very turbulent liquids allows, for example, the rapid flow of very viscous molten lava.

4.11 Bernoulli's Equation

Rather than consider the actual velocity of a liquid flowing through a tube, which is dependent on the viscosity of the liquid, there is a different approach to the problem of flowing liquids. This is simply a statement of the fact that what goes in one end of a tube must come out of the other, and includes both volume and energy considerations. The flow must, however, be streamlined, otherwise the unpredictable effects of turbulence would make the problem and its solution too complex, and we must consider energy losses due to viscosity as being small. Such an approach has the advantage that it depends on the cross-sections of the ends of the tube and is in no way concerned with the shape of the tube between the ends.

Consider the tube AB in Figure 4.14 in which liquid is entering at A and leaving at B. The tube is of no particular cross-sectional or longitudinal shape and may be inclined to the horizontal, as shown. An elemental volume of liquid at A will be moving with a velocity v_A and occupy a length δs_A. When this elemental volume reaches end B, it will be moving with a velocity v_B and will occupy a length δs_B. Let us now calculate the work required to push the liquid through the tube. To cause the

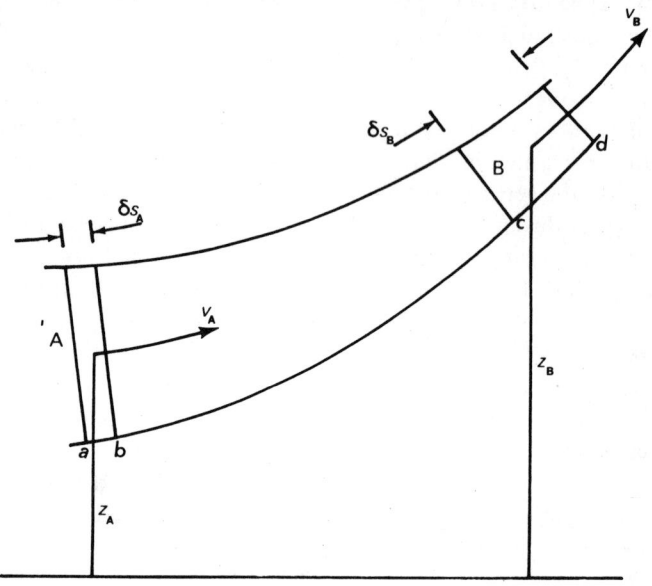

Figure 4.14.   Illustration for the derivation of Bernoulli's equation

liquid to flow, that is, to cause the elemental volume to move from A to B, a force F_s must act on the face a of the element, where s is measured along the direction of flow, along a streamline. The work done then in moving the elemental volume through the tube, that is, moving the face a at A to the position of face c at B is

$$\int_a^c F_s \, ds$$

If A is the cross-sectional area of the tube at any point and p is the corresponding pressure at that point, then we can write that

$$\int_a^c F_s \, ds = \int_a^c pA \, ds$$

$$= \int_a^b pA \, ds + \int_b^c pA \, ds \qquad (4.50)$$

150

However, since there is a pressure acting everywhere within the liquid, the liquid itself will do work to resist the flow. Thus the force F_s acting against the direction of flow may be imagined as acting on the face b at A and resisting its movement to its corresponding position d at B. Then, as before,

$$\int_b^d F_s \, ds = \int_b^d pA \, ds$$

$$= \int_b^c pA \, ds + \int_c^d pA \, ds \qquad (4.51)$$

Hence the net work done in moving the elemental volume of liquid through the tube, from A to B, is

$$\int_a^c F_s \, ds - \int_b^d F_s \, ds = \int_a^b pA \, ds + \int_b^c pA \, ds - \int_b^c pA \, ds - \int_c^d pA \, ds$$

$$= \int_a^b pA \, ds - \int_c^d pA \, ds \qquad (4.52)$$

If we consider sufficiently small elements of volume, then the pressure and cross-sectional area may be considered constant over the region a to b, and similarly from c to d. Equation 4.52 may thus be written as

$$\text{net work} = p_A A_A \, \delta s_A - p_B A_B \, \delta s_B$$

$$= (p_A - p_B) V \qquad (4.53)$$

where the suffixes indicate values at the respective ends A and B of the tube and V is the volume of the element, that is, $V = A_A \, \delta s_A = A_B \, \delta s_B$. Since also $V = m/\rho$, where m is the mass of the element and ρ is the density of liquid, equation 4.53 in turn becomes

$$\text{net work} = (p_A - p_B)\frac{m}{\rho} \qquad (4.54)$$

However, we know that the net work done is equal to the change in kinetic and potential energies produced, so

$$(p_A - p_B)\frac{m}{\rho} = \left(\frac{1}{2}mv_B{}^2 - \frac{1}{2}mv_A{}^2\right) + (mgz_B - mgz_A) \qquad (4.55)$$

where the change in potential energy $(mgz_B - mgz_A)$ is due to the change in elevations z_A and z_B of the mass m of liquid between the ends of the tube, at A and B. From equation 4.55,

$$p_A - p_B = \frac{1}{2} \rho (v_B^2 - v_A^2) + \rho g(z_B - z_A) \qquad (4.56)$$

that is,

$$p_A + \frac{1}{2} \rho v_A^2 + \rho gz_A = p_B + \frac{1}{2} \rho v_B^2 + \rho gz_B \qquad (4.57)$$

In other words,

$$p + \frac{1}{2} \rho v^2 + \rho gz = \text{const.} \qquad (4.58)$$

along a streamline. This is *Bernoulli's equation*. It should be remembered that it neglects energy losses due to viscosity and any losses due to turbulence.

A briefer derivation of this equation is simply to state that the sum of the kinetic and potential energies of the volume element at A, plus the work *done on* it in a time δt as it enters the tube, is equal to the sum of the kinetic and potential energies at B, plus the work *done by* the volume element in a time δt as it leaves the tube. If in this time δt a mass m flows across a cross-section at A and occupies a length δs_A of the tube, and similarly in the same time the same mass occupies a length δs_B at B, then

$$\frac{1}{2} mv_A^2 + mgz_A + p_A A_A \delta s_A = \frac{1}{2} mv_B^2 + mgz_B + p_B A_B \delta s_B \qquad (4.59)$$

Equations 4.57 and 4.58 then follow as before.

In particular it should be noted that the equation of hydrostatics is a special case of Bernoulli's theorem. For a volume of liquid in which the velocity is everywhere zero, equation 4.56 becomes

$$p_A - p_B = \rho g(z_B - z_A) \qquad (4.60)$$

where $p_A - p_B$ is the pressure difference between two points a vertical distance $z_B - z_A$ apart.

As an example of the use of Bernoulli's equation, we may consider the problem of the rate at which water would flow through a long tube AB whose axis is horizontal, whose radius at A is 3 cm and at B is 2 cm, and where the pressure difference between A and B is equivalent to 10 cm of water. Since the axis is horizontal, $z_A = z_B$ and equation 4.56 simplifies to

$$p_A - p_B = h\rho g = \frac{1}{2}\rho\,(v_B{}^2 - v_A{}^2)$$

that is,

$$v_B{}^2 - v_A{}^2 = 2hg$$

where h is the head of water, g the intensity of gravity, and v_A and v_B are the velocities of the flow at A and B respectively. Substituting values,

$$v_B{}^2 - v_A{}^2 = 2 \times 0 \cdot 1 \times 9 \cdot 81$$

Also, the volume passing through the tube per second is $\pi r^2 v$, that is,

$$\pi \times 0 \cdot 03^2\, v_A = \pi \times 0 \cdot 02^2\, v_B$$

so

$$v_A = \frac{0 \cdot 02^2}{0 \cdot 03^2}\, v_B = \frac{4}{9}\, v_B$$

Therefore

$$v_B{}^2 \left[1 - \left(\frac{4}{9}\right)^2\right] = 2 \times 0 \cdot 1 \times 9 \cdot 81$$

from which

$$v_B = 1 \cdot 563 \text{ m s}^{-1}$$

This is the velocity of the liquid at the end B. Therefore the volume leaving the end B per second, which is of course the volume per second passing right through the tube, is $\pi r_B{}^2 v_B$, that is,

$$\text{volume per second} = 3 \cdot 14 \times 0 \cdot 02^2 \times 1 \cdot 563$$
$$= 1 \cdot 96 \times 10^{-3} \text{ m}^3 \text{ s}^{-1}$$

4.12 Applications of Bernoulli's Equation

From equation 4.58 it can be seen that, whenever the velocity of flow is increased, the pressure is decreased. Thus in a hurricane the roof of a building may be sucked off, rather than blown off, owing to the increased velocity of flow, as shown in Figure 4.15(a); therefore it is often recommended that doors and windows are left open in an attempt to equalise the pressures. Similarly, delicate small components, such as chips to be used in semiconductor devices, may be lifted by an arrangement shown in Figure 4.15(b). To the end of a tube passing compressed

153

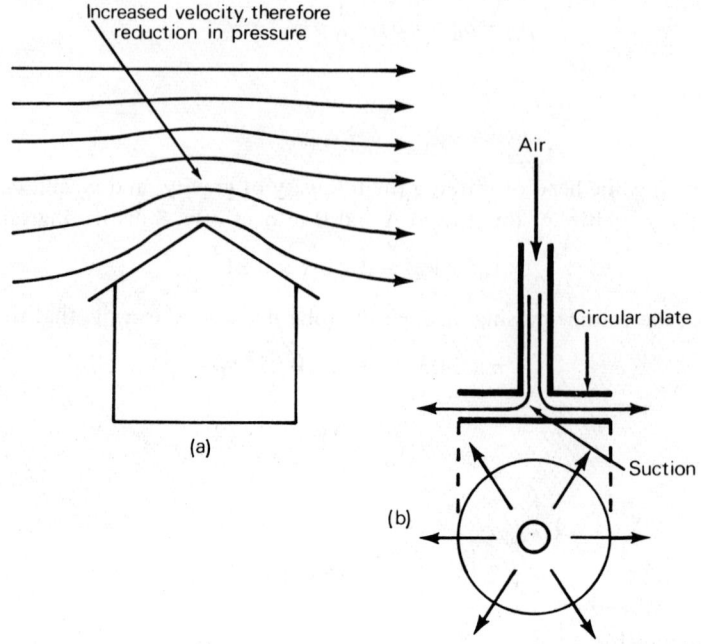

Increased velocity, therefore
reduction in pressure

Air

Circular plate

Suction

(a)

(b)

Figure 4.15.   Applications of Bernoulli's equation

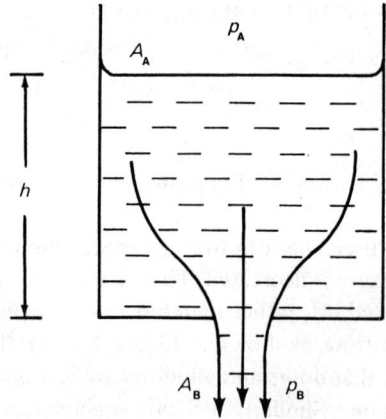

A_A
p_A

h

A_B p_B

Figure 4.16.   Illustration for the derivation of Torricelli's theorem

air is fitted a circular plate. As this plate is brought close to the flat object to be lifted, the air flow will become radial in the parallel gap, so the flow velocity will be greatest near the centre with a consequent reduction of pressure, which will become approximately atmospheric at the periphery. The flat object will be lifted owing to the reduced pressure in the gap but, since the air must continue to flow across this gap, the object cannot be brought into contact with the end of the air pipe and its circular plate, when it may possibly become scratched. This is an obvious advantage over a simple vacuum-operated lifting arrangement.

Let us now consider a tank containing a liquid to a height h, as shown in Figure 4.16, where the top of the tank of area A_A is open to the atmosphere. Let there also be a hole of area A_B in the bottom of the tank. Since the tank is open to the atmosphere, $p_A = p_B = p$, the atmospheric pressure, and therefore $p_A - p_B = 0$. We may now apply Bernoulli's equation, in the form of equation 4.56, since this tank and its hole may be likened to the tube of Figure 4.14. Thus

$$0 = \frac{1}{2} \rho (v_B{}^2 - v_A{}^2) + \rho g (z_B - z_A)$$

that is,

$$v_B{}^2 - v_A{}^2 = 2gh \tag{4.61}$$

where v_B is the velocity of efflux of the liquid from the hole, v_A is the velocity at which the liquid level is falling in the tank, and $h = z_A - z_B$ is the height of liquid in the tank. If the hole is small so that $A_B \ll A_A$, then the velocity v_A is very small in comparison with the velocity v_B through the hole. Therefore equation 4.61 can be written as

$$v_B = \sqrt{(2gh)} \tag{4.62}$$

That is, the velocity of efflux v_B is the same as that acquired by any body that falls freely under gravity through a height h. This is *Torricelli's theorem*. Equation 4.62 is equally applicable to a tank with a small hole in the side, where h is still the height of the liquid level above the hole.

Bernoulli's equation may also be applied to flow through a *Venturi tube*. This is a tube in which there is a constriction or throat, as shown in Figure 4.17. The design of the constriction is such that streamlined flow is maintained – no turbulence is introduced. Let p_A and v_A be the pressure and velocity respectively in the clear tube, and similarly in the constriction, p_B and v_B. If Q is the volume of liquid passing through

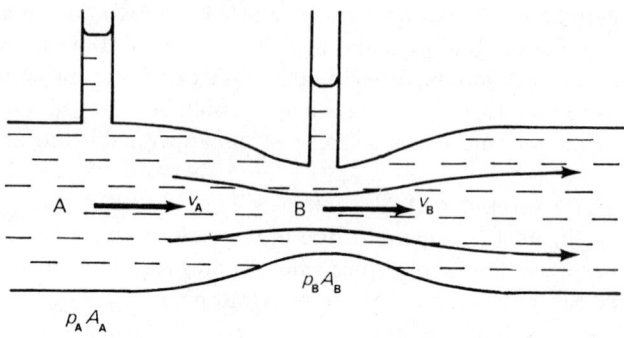

Figure 4.17. A Venturi tube, showing increased flow rate and
reduced pressure at the constriction

the tube in unit time, then $v_A = Q/A_A$ and $v_B = Q/A_B$. Substituting in equation 4.56,

$$p_A - p_B = \frac{1}{2}\rho Q^2 \left(\frac{1}{A_B{}^2} - \frac{1}{A_A{}^2}\right) \tag{4.63}$$

where ρ is the density of the liquid and the axis of the tube is horizontal. Hence the volume of liquid flowing through the tube in unit time is

$$Q = kA_A A_B \sqrt{\frac{2(p_A - p_B)}{\rho(A_A{}^2 - A_B{}^2)}} \tag{4.64}$$

where the coefficient k must be determined experimentally and is introduced to take account of losses due to viscous drag and eddy currents occurring in a particular tube. The pressure difference $p_A - p_B = \rho g h$ is determined by means of the vertical tubes as shown in the figure. An instrument using this principle to measure the flow rate is a *Venturi meter.* As well as this meter for measuring flow rates in tubes, the *Pitot tube* is also used for measuring flow rates, but not necessarily in tubes. It is used in measuring, for example, the flow of water past ships and air past aircraft and as such it measures the velocity of the craft relative to its surroundings.

Figure 4.18 shows a liquid flowing past a horizontal surface. Protruding through the surface is a Pitot tube in the form of a manometer

156

connected to a probe with an opening in the upstream end. In Figure 4.18(b) the flow lines are shown in the region of the probe nozzle. The point at which the horizontal velocity becomes zero in Figure 4.18(b) is termed the *stagnation point*. Labelling the two liquid regions A and

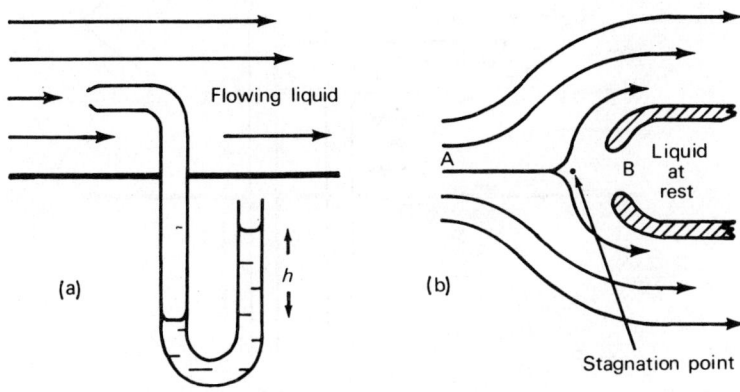

Figure 4.18. A Pitot tube for measuring the velocity of a flowing liquid

B as in the figure, where A is sufficiently far from the nozzle for the flow lines to be unaffected, and applying equation 4.57,

$$p_A + \frac{1}{2}\rho v_A{}^2 = p_B \qquad (4.65)$$

since the liquid is at rest within the probe and hence $v_B = 0$. The manometer measures the pressure p_B and, if the *static pressure* p_A is measured, the velocity v_A of the liquid relative to the Pitot tube may be calculated. The manometer in fact does not measure the pressure p_B directly but its excess over the atmospheric pressure p_{atmos}. That is,

$$p_{atmos} - p_B = g\rho_m h \qquad (4.66)$$

where ρ_m is the density of the manometric liquid. Both $p_A + \frac{1}{2}\rho v_A{}^2$ and $\frac{1}{2}\rho v_A{}^2$ are also loosely referred to as the *dynamic pressure*.

A convenient modification of the simple Pitot tube is the *Prandtl tube* shown in Figure 4.19. This arrangement is also commonly referred to as a 'Pitot' tube. The pressure in the nozzle, which should be ellipsoidal in shape, is p_B, given by equation 4.65. By having a ring of holes in the side of the probe, parallel to the liquid flow lines, the static pressure p_A

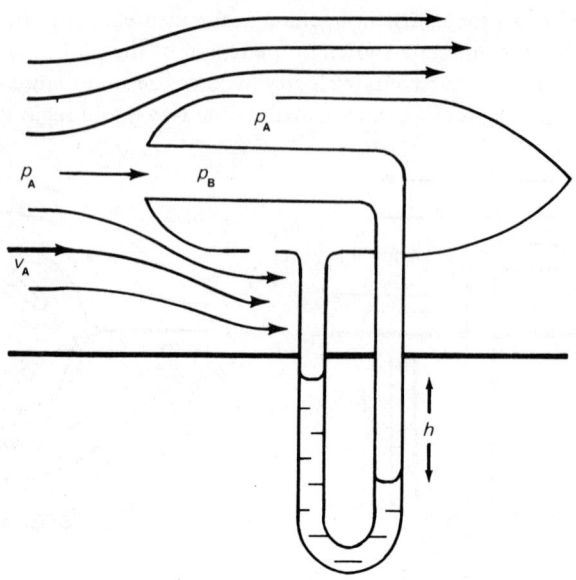

Figure 4.19. Arrangement of a Prandtl tube for measuring flow rate

is also recorded. The manometer now gives the differential pressure $p_B - p_A$, that is,

$$p_B - p_A = \frac{1}{2} \rho \, v_A^2 = g \rho_m h \qquad (4.67)$$

as in equation 4.65. For high velocities in gases it may be necessary to take account of changes in the density ρ.

4.13 Dynamic Lift

We have been much concerned in this chapter with the movement of liquids along streamlines. It is therefore not out of place to discuss how lift can result from the asymmetry of streamlines. If we were to discuss a lighter-than-air balloon, we would be concerned with *static lift*, which is due to the change in atmospheric pressure with elevation and a resulting greater pressure on the lower half of the balloon than on the upper half. *Dynamic lift*, however, results from the flow over a surface, for example air over a moving aircraft's wing. By deflecting the streamlines, a greater velocity of flow can be achieved over the upper surface, as

158

shown in Figure 4.20, with a consequent reduction in pressure and resulting lift.

A similar lift can be obtained with a spinning body moving through a fluid. For example, Figure 4.21(a) represents symmetrical streamlines flowing round a stationary cylinder. If now we consider a spinning

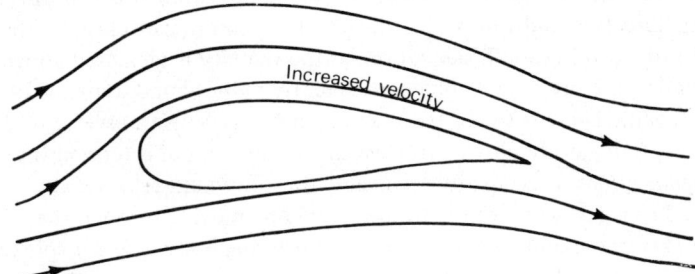

Figure 4.20.   Dynamic lift resulting from an increased velocity of flow over an airfoil

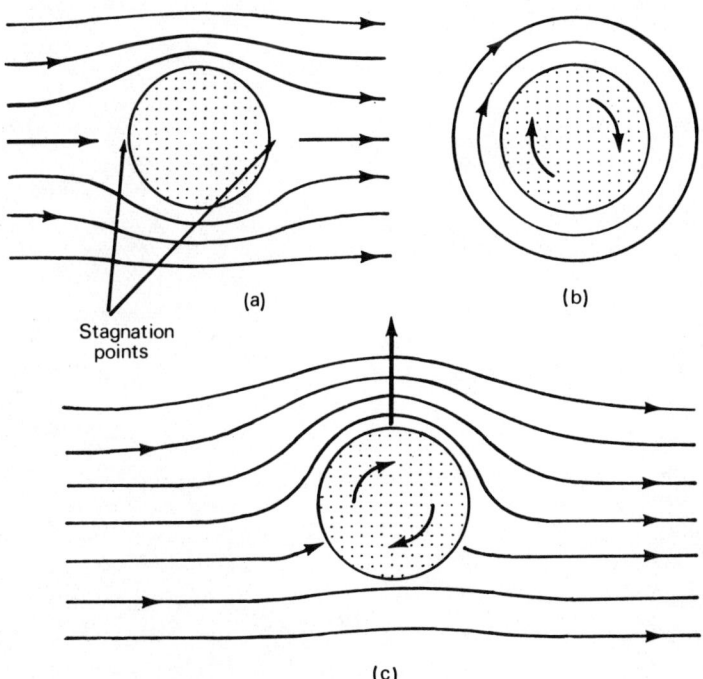

(a)

(b)

Stagnation
points

(c)

Figure 4.21.   The transverse force on a rotating body moving relative to a fluid medium

cylinder in a stationary air stream, the air in the neighbourhood of the cylinder will be dragged round as a result of the viscosity of the air and a circulating flow pattern will result, as shown in Figure 4.21(b). By combining these motions, a spinning cylinder producing a circulating flow pattern and a moving air stream, we obtain the streamlines shown in Figure 4.21(c), since the velocities above the cylinder are in the same direction and add, whilst those below tend to cancel each other. According to Bernoulli's equation, as the velocity is increased above the cylinder, the pressure will be consequently reduced and dynamic lift will result. There is thus a force acting upwards. This transverse force on a rotating body moving relative to a fluid medium is termed the *Magnus effect*. It is equally applicable for movement relative to a liquid or a gas, whether the rotating body be moving through the liquid or the liquid be flowing past the rotating body, and for spheres as well as cylinders. A common example is the curved path taken by a sliced golf or tennis ball.

Chapter Five

VISCOSITY OF GASES AND NON-NEWTONIAN FLUIDS

5.1 Introduction

In the previous chapter we were concerned with the flow of liquids and bodies moving through liquids but, in particular, we were concerned with flow that was in accordance with Newton's law of viscosity (equation 4.1). This is concerned with streamline flow. We also dealt only with liquids that are assumed incompressible. In this chapter we shall show how to apply Newton's equation to gases, which are compressible. First, however, we shall examine how the viscosity of gases may be deduced theoretically from a discussion of the motion of the individual molecules of a gas and then go on to show that methods of measurement applicable to liquids can in general be adapted for gases.

Having dealt with the viscosity of gases, we shall proceed to discuss that group of fluids that does not obey Newton's equation even for slow flow rates. These liquids have probably the greatest industrial importance but unfortunately do not lend themselves to a theoretical treatment. Equations that have been derived for the flow of these liquids have only very limited application under specialised conditions and therefore we must content ourselves with a more general treatment and discussion.

5.2 Viscosity and the Kinetic Theory of Gases

To be able to deduce any useful equations for gases, we shall simplify the problem slightly by assuming a perfect gas. By this we mean a gas in which the molecules are assumed to be infinitely small spheres so that their volume may be neglected in comparison with the total volume of gas. We shall also assume that the molecules exert only negligible force on each other so that interaction only occurs when molecules collide with each other or the walls of the containing vessel. These collisions, we shall assume, are on the average perfectly elastic; if this were not so,

161

the total kinetic energy of the gas molecules would soon fall. Owing to thermal energy, the molecules will always be in a state of violent agitation, moving in random directions and, since there are some 10^{25} molecules in a cubic metre of gas at normal atmospheric pressure and temperature, there will be a large number of collisions occurring at any time. It is these collisions with the walls of the containing vessel that are manifested as a pressure.

Since so many collisions are involved, the average distance that a molecule of the gas travels between successive collisions is an important concept of the kinetic theory. This distance is termed the *mean free path length, l.* We can relate this to the diameter d of a gas molecule and to the number density n, that is, the number of molecules in unit volume, in the following way. Between collisions we may assume a gas molecule travels in a straight line since we have said the molecules exert only negligible force on each other. Now, if a molecule is moving along a path as shown in Figure 5.1, a collision will take place with

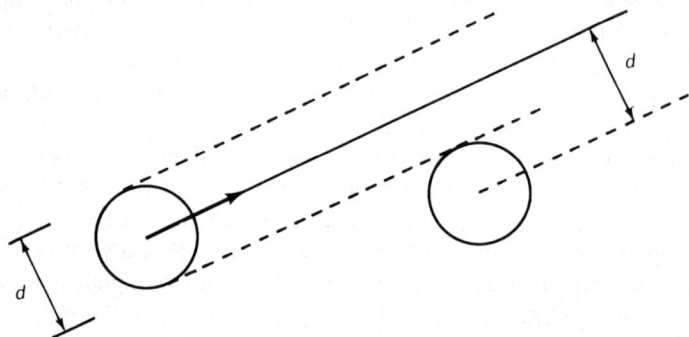

*Figure 5.1.   Illustration for the calculation of mean free path length
of a molecule*

another molecule if its centre is less than d from the line of path taken by the centre of the moving molecule. If the average speed of a moving molecule at the appropriate temperature is $\bar{c}$, on average a molecule will travel a distance $\bar{c}t$ in a time t. Note that we are referring to speed, a scalar quantity, rather than velocity, a vector quantity, because we are not here concerned with any special direction of motion. The number of collisions that the moving molecule will have with other molecules will thus be equal to the number of molecules whose centres are within a cylindrical volume $\pi d^2 \bar{c}t$, that is, $n\pi d^2 \bar{c}t$, where n is the number of molecules per unit volume. This is the number of collisions occurring in

162

a time t and, since in this time a molecule travels a distance $\bar{c}t$, the mean distance it travels between successive collisions, the mean free path length, is

$$l = \frac{\bar{c}t}{n\pi d^2 \bar{c}t} = \frac{1}{n\pi d^2} \qquad (5.1)$$

We have, however, regarded only one molecule as moving, with the others stationary. If we take into account that all the molecules are in fact in motion, it may be shown that the length of the mean free path is more correctly

$$l = \frac{1}{n\pi d^2 \sqrt{2}}$$

$$= \frac{1}{n \sigma \sqrt{2}} \qquad (5.2)$$

where $\sigma = \pi d^2$ is termed the *collision cross-section*. In practice, of course, the mean free path length is a very difficult quantity to measure since the diameter of a molecule has no exact physical meaning. It is nevertheless of importance in the kinetic theory of gases to appreciate that a mean free path does exist even if its length is not readily measurable.

Let us now return to the problem of viscosity and its implication in the kinetic theory of gases. We have already discussed in the previous chapter how viscosity arises from the relative motion of layers of liquid and the friction between them. The same applies to a gas moving with a streamlined motion.

A layer of gas at some distance z from a fixed plane xy and parallel to it is moving with some velocity v in the x direction, as shown in Figure 5.2. A parallel layer at a distance $z + \delta z$ is moving with a velocity $v + \delta v$, as shown. It must be remembered that these are over-all velocities superimposed on the random thermal velocities which, of course, have no preferred direction of drift. Nevertheless, we must still consider the random thermal motion of the individual molecules, some of which will be moving transversely across the layers. Those moving from a slow layer to the adjacent faster-moving layer will tend to retard the drift motion of the molecules in this layer and, conversely, those molecules moving into a slower moving layer will tend to increase its drift velocity. Thus there will be a net transfer of momentum from the faster to the slower moving layers which will tend to equalise their velocities. The rate of transfer of momentum, by Newton's second law of motion, will constitute a force which in this case may be regarded as a frictional force giving rise to viscosity. We now have to apply the concepts of the kinetic theory of gases to derive an expression for this rate of transfer of momentum.

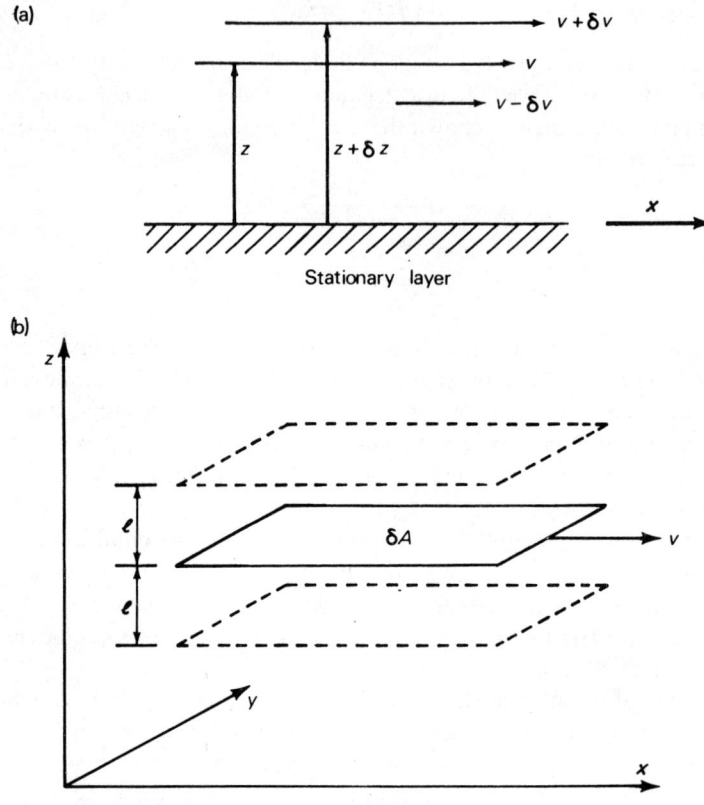

Figure 5.2.   Illustrating the flow of gas over a surface

Returning again to Figure 5.2, let us consider an element of area δA in the layer moving with a drift velocity v in the x direction, as shown in Figure 5.2(b). If we were to imagine a cube of unit volume in the gas, then, since all the molecules are moving in completely random directions, we may assume that at any instant one-sixth of the total number of molecules in the cube, that is, $n/6$, will be travelling towards any given one of its faces. Thus, if n is the number of molecules in unit volume and $\bar{c}$ the mean speed of the molecules, then the number

164

of molecules passing in one direction across the area δA in unit time will be

$$n' = \frac{n \delta A}{6} \bar{c} \qquad (5.3)$$

Each of these molecules, however, will have travelled, on average, a distance l, the mean free path length, since its last collision. Also a molecule of mass m moving with a drift velocity v in the x direction when it is distance z from the fixed plane has a momentum mv, and we may assume that, at its last collision before crossing through the area δA, each molecule acquired a drift velocity in the x direction. This acquired drift velocity will depend on the distance from the fixed surface at which the collision took place. Molecules that cross the elemental area δA from above, since they have a greater velocity in the x direction, will carry over a greater momentum than those crossing from below. The momentum carried over from above by each molecule is thus

$$m \left(v + l \frac{dv}{dz} \right)$$

and that carried over from below is

$$m \left(v - l \frac{dv}{dz} \right)$$

The rate at which linear momentum p is transported across the elemental area δA by the n' molecules that cross in unit time is thus

$$\frac{dp}{dt} = \frac{n \bar{c} \delta A}{6} \left[m \left(v + l \frac{dv}{dz} \right) - m \left(v - l \frac{dv}{dz} \right) \right]$$

$$= \frac{nml\bar{c}\delta A}{3} \frac{dv}{dz} \qquad (5.4)$$

where n is the number of molecules in unit volume, each of mass m and moving with a mean speed $\bar{c}$ due to thermal agitation. This rate of change of momentum, as we have already said, constitutes a force according to Newton's second law of motion. Since it is the change in momentum that leads to viscosity in this case, the resulting force can be regarded as the viscous drag force. That is,

$$F = \frac{nml\bar{c}\delta A}{3} \frac{dv}{dz} = \eta \, \delta A \frac{dv}{dz} \qquad (5.5)$$

165

from Newton's equation for viscosity (equation 4.1). Thus the coefficient of viscosity is given by

$$\eta = \frac{nml\bar{c}}{3} \tag{5.6}$$

Since also $mn = \rho$, the density of the gas,

$$\eta = \frac{\rho l\bar{c}}{3} \tag{5.7}$$

where l, the mean free path length, is given by equation 5.2. Alternatively, η may be written as

$$\eta = \frac{m\bar{c}}{3\sigma\sqrt{2}} \tag{5.8}$$

from equations 5.6 and 5.2.

If now the pressure of the gas is increased, then the density will also increase but the mean free path length will decrease, so, from equation 5.7 or directly from equation 5.8, it is seen that the viscosity of the gas is independent of its pressure. This fact is borne out by experiment, except at very low pressures where the mean free path length becomes of the order of the dimensions of the containing vessel and at very high pressures where the forces between the molecules become of importance. Thus this is an important confirmation of the correctness of the principles underlying the kinetic theory.

It has also been shown that the mean speed $\bar{c}$ of the molecules, due to thermal agitation, varies as the square root of the absolute temperature. Consequently, from equation 5.8, the viscosity also varies as the square root of the absolute temperature – a fact again borne out by experiment.

There is also a close association, perhaps surprisingly, between the viscosity of a gas and its thermal conductivity. Let us consider a gas trapped between two parallel plates of large area, so that we may neglect edge effects, with the upper plate maintained at a higher temperature than the lower. This way round, convection effects are minimised. The situation is thus as shown in Figure 5.3, where layers of gas are at the temperatures shown. This figure may be compared with Figure 5.2, although there is now a temperature gradient whereas in the earlier figure there was a velocity gradient. Transport of molecules across an elemental area δA is still occurring, but this time we are considering the molecules as carrying thermal energy rather than momentum. The thermal energy of a molecule may be written as $E = mc_v T$, where

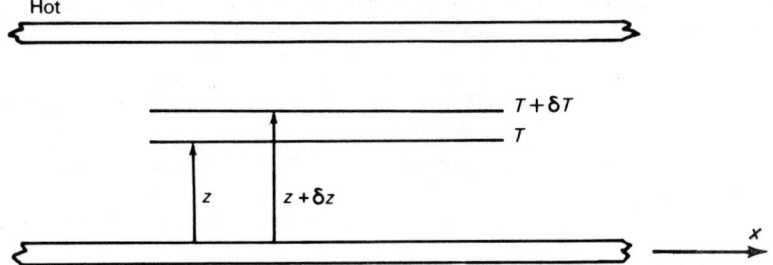

Figure 5.3. *Illustrating the thermal conductivity of a gas between two plates at different temperatures*

m is the mass of a molecule and c_v and T are the specific heat at constant volume and the absolute temperature of the gas respectively. Thus, by analogy with equation 5.4, the rate of transport of thermal energy across the elemental area δA is

$$\frac{dE}{dt} = \frac{nl\bar{c}\delta A}{3}\left(mc_v\frac{dT}{dz}\right) = \delta Q \qquad (5.9)$$

where δQ is the amount of heat transferred across the area δA in unit time. But by definition

$$\delta Q = \lambda\,\delta A\,\frac{dT}{dz} \qquad (5.10)$$

where λ is the thermal conductivity and dT/dz is the temperature gradient. Hence, from equations 5.9 and 5.10,

$$\lambda = \frac{nml\bar{c}c_v}{3} \qquad (5.11)$$

By substituting from equation 5.6,

$$\lambda = \eta c_v \qquad (5.12)$$

A fuller treatment, taking into account that rotational and translational energies are transferred and that the molecules do not all travel the mean free path length before crossing the area δA, gives a numerical correcting factor to equation 5.12, that is,

$$\lambda = f\eta c_v \qquad (5.13)$$

where f varies from unity to $\frac{5}{2}$ depending on the gas.

The viscosities of some typical gases are given in Table 5.1.

Table 5.1.VISCOSITIES OF SOME GASES AT 20°C AND 1 atm
(UNLESS OTHERWISE INDICATED)

Gas	Dynamic viscosity, η (10^{-7} N s m^{-2} or micropoise)	Kinematic viscosity, ν (10^{-10} m^2 s^{-1} or microstokes)
Air	$\begin{cases} 171(0°C) \\ 182 \\ 191(40°C) \end{cases}$	$\begin{cases} 132(0°C) \\ 151 \\ 169(40°C) \end{cases}$
Argon	223	134
Carbon dioxide	147	80
Chlorine	133	45
Helium	196	1180
Hydrogen	89	1060
Methane	110	165
Nitrogen	176	151
Oxygen	203	153
Propane	80	44

5.3 Gas Flow through a Tube

In the previous chapter we discussed the Hagen—Poiseuille equation
for the streamlined flow of a liquid through a tube. This considered the
liquid to be incompressible, which a gas most certainly is not. However,
the equation can be modified to take the compressible nature of the
fluid into account. In deriving the Hagen—Poiseuille equation we considered the velocity of flow at different distances from the axis of the
tube and treated the volume of liquid as constant across any cross-
section. However, for a gas we must make use of the fact that equal
masses of gas rather than equal volumes pass each cross-section in a
given time. That is, ρv must be constant at every point along a line at a
given distance from the axis of the tube, where ρ is the density of the
gas at the particular point. Since the density of the gas will decrease
towards the low-pressure end of the tube, that is, the exit, the velocity
must increase at the same time but, since also the density will vary as
the pressure, then ρv must also be constant at every point along this
same line, where p is the pressure at the particular point. We can also
consider this pressure to be constant across a particular cross-section,
that is, p is independent of r, the radius of the tube.

Now, returning to the Hagen—Poiseuille equation (equation 4.20),
that is,

$$Q = \frac{\pi(p_1 - p_2)r^4}{8\eta l}$$

where Q is the volume of liquid flowing through the tube of length l and radius r under a pressure differential $p_1 - p_2$, we can rewrite this as

$$Q = -\frac{\pi r^4}{8\eta}\frac{\mathrm{d}p}{\mathrm{d}x} \qquad (5.14)$$

where $\mathrm{d}p/\mathrm{d}x$ is the rate of change of pressure with distance along the tube. The negative sign arises since p decreases as x increases. Taking into account that for a gas ρv is constant across any cross-section, pQ is also constant across any cross-section, where Q is the volume of gas at a pressure p passing that cross-section in one second.

If p_1 is the pressure of the gas at the inlet of the tube and Q_1 the volume of gas entering in one second at this pressure, then, from equation 5.14,

$$p_1 Q_1 = pQ = -\frac{\pi r^4}{8\eta} p \frac{\mathrm{d}p}{\mathrm{d}x} \qquad (5.15)$$

Thus, for a tube of length l and with a gas pressure at the outlet of p_2,

$$\int_0^l p_1 Q_1 \, \mathrm{d}x = -\frac{\pi r^4}{8\eta}\int_{p_1}^{p_2} p \, \mathrm{d}p \qquad (5.16)$$

That is,

$$p_1 Q_1 l = -\frac{\pi r^4}{8\eta}\left(\frac{p_2{}^2 - p_1{}^2}{2}\right)$$

and

$$Q_1 = \frac{\pi r^4}{16\eta l}\left(\frac{p_1{}^2 - p_2{}^2}{p_1}\right) \qquad (5.17)$$

where Q_1 is the volume of gas entering the tube per second at the pressure p_1 and, if Q_2 is the volume leaving per second at a pressure p_2, then Q_2 is given by $p_1 Q_1 = p_2 Q_2$. Equation 5.17 is due to Meyer.

When considering the flow of a liquid through a tube, we assumed that the layer adjacent to the tube wall was stationary. Experiments show that this cannot strictly be assumed to be the case for a gas and there is some slipping of this outer layer relative to the tube wall. The gas flows as if the radius of the tube were greater by an amount approximately equal to the length of the mean free path of the gas at the particular pressure concerned. Thus the gas flows as if the radius were $r + \beta$, where β is a function of the mean free path length and may

169

be taken as a constant for a particular gas; β is termed the *slipping coefficient*. Then

$$(r + \beta)^4 = r^4 \left(1 + \frac{\beta}{r}\right)^4$$

$$\simeq r^4 \left(1 + \frac{4\beta}{r}\right)$$

Hence, substituting in equation 5.17,

$$Q_1 = \frac{\pi r^4}{16\eta l} \left(\frac{p_1{}^2 - p_2{}^2}{p_1}\right) \left(1 + \frac{4\beta}{r}\right) \tag{5.18}$$

where β is the slipping coefficient. Again the Hagenbach corrections apply, as for the liquid case discussed in Section 4.3, so the viscosity of the gas is given by

$$\eta = \frac{\pi^2 r^4 (p_1{}^2 - p_2{}^2)(1 + 4\beta/r) - 2p_1 Q_1{}^2 [m + \log_e (p_1/p_2)]}{16\pi p_1 Q_1 l} \tag{5.19}$$

where m is a correcting multiplying constant approximately equal to unity. This equation may be contrasted with that for an incompressible liquid (equation 4.30).

5.4 Use of a Rotating Cylinder Viscometer with Gases

It should be noted that in general any method suitable for measuring the viscosity of a liquid may be adapted for the measurement of the viscosity of a gas. The rotating cylinder viscometer is no exception. We have already shown in the previous chapter (equation 4.36) that the couple T that must be exerted on the inner cylinder to cause it to rotate at a constant angular velocity ω is

$$T = 4\pi\eta l\omega \left(\frac{r_1{}^2 r_2{}^2}{r_2{}^2 - r_1{}^2}\right) \tag{5.20}$$

This equation neglects the 'end effects'. A more convenient arrangement for a gas than that described in Section 4.5 is to rotate the inner cylinder as before and support the outer cylinder by a torsion wire. Rotating the inner cylinder at a constant angular velocity drags the outer cylinder round by an amount depending on the viscosity of the gas. Such an arrangement is shown diagrammatically in Figure 5.4. The inner cylinder is driven remotely by an electric motor working through a magnetic clutch, whilst the outer cylinder is suspended at three points and fixed to a torsion wire. Deflection of this outer cylinder is measured by the usual optical lever principle utilising a mirror fixed at the bottom end of the torsion suspension. This outer cylinder is also shorter than the inner cylinder so that guard rings may be used to eliminate end effects.

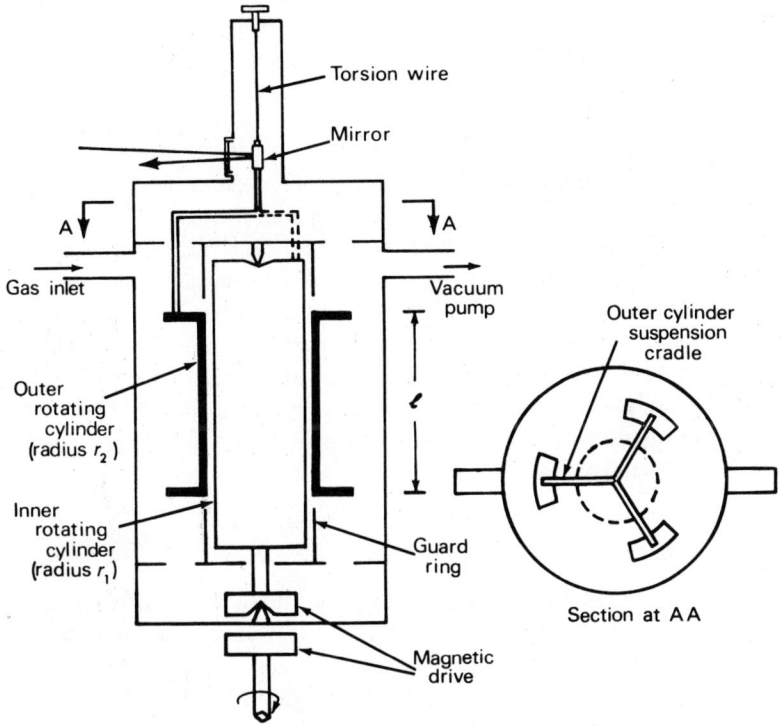

Figure 5.4.   A rotating cylinder viscometer for use with gases

Finally the whole apparatus may be evacuated and refilled with the gas under test. The apparatus, of course, is also maintained at a fixed temperature.

When the inner cylinder is rotating at a constant angular velocity, the outer cylinder will be rotated by some angle θ owing to the drag of the gas. In practice, the inner cylinder should be rotated in both directions in turn so that θ is half the total deflection obtained. The restoring couple exerted on the outer cylinder by the torsion wire is thus $k\theta$, where k is the couple required to give unit twist to the suspension. To determine the torsional constant k of the wire, various simple-shaped

171

bodies are suspended on the wire and allowed to oscillate in vacuum. The period of oscillation is $2\pi\sqrt{(I/k)}$, where I is the moment of inertia of the suspended body, so k may be determined. It should also be confirmed by using different length outer cylinders with corresponding length guard rings that the length l of the cylinder occurring in equation 5.20 is in fact the physical length of the outer cylinder.

5.5 Non-Newtonian Flow

If a shear force is applied across a solid, it will be deformed in a manner which we have already discussed, for example in Chapter 1. The rate at which the shear force is applied, in general, has negligible effect. Thus the shear modulus may be assumed to be a constant for a particular material. Similarly, for the liquids we have dealt with so far, the coefficient of viscosity is a constant for a particular liquid at a given temperature. Departures from this occur as the velocity of flow is increased and turbulence becomes of importance.

From equation 4.1 for a liquid obeying Newton's law of viscous flow, the viscosity at a given temperature is given by

$$\eta = \frac{F/A}{dv_x/dz} = \frac{\text{shear stress, } \tau}{\text{rate of shear, } S} \qquad (5.21)$$

However, for many liquids there is no constant coefficient of viscosity. For such liquids, termed non-Newtonian, it may vary by many orders of magnitude with changing rate of shear.

In Figure 5.5 is plotted the shearing stress against the corresponding rate of shear for various types of liquid. Such a curve is known as a *consistency curve*. Curve A is in fact a straight line, giving a constant coefficient of viscosity, and therefore is the curve associated with a Newtonian liquid. For some liquids the rate of shear increases faster than the rate of increase of the shearing stress. These are termed *pseudoplastic* or *shear-thinning liquids* and are represented by curve B, which has no linear portion. In other words, there is no coefficient of viscosity in the usual sense. We can only really speak of an *apparent viscosity*, that is, the ratio of the shearing stress to rate of shear at a given value of, say, the rate of shear. Curve C is the converse of curve B and is for a liquid whose rate of shear increases slower than the rate at which the shearing stress is increasing. This is termed a *dilatant* or *shear-thickening liquid*. Both curves B and C can be represented by a power law to a first approximation, that is, straight lines would be obtained if log rate of shear were plotted against log shearing stress. Thus log apparent viscosity would also be proportional to either log rate of shear or log shearing stress, at least to a first approximation.

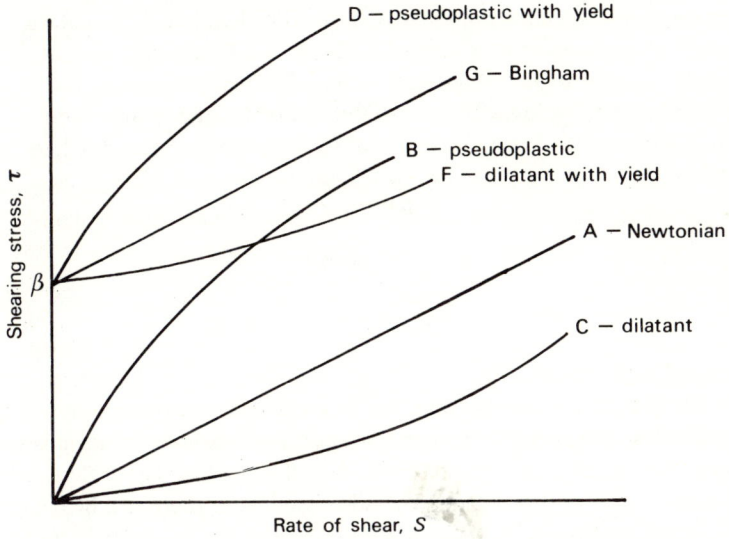

Figure 5.5.   Flow curves to show the properties of some non-Newtonian liquids

Pseudoplastic liquids are probably the most common type having industrial applications. It is seen from Figure 5.5 that the apparent viscosity $\eta_{app} = \tau/S$ is high for low shearing stresses but falls as the shearing stress increases. This fall is due to some form of chemical bonding which breaks down at an increasing rate as the shearing stress is increased. If this bonding is very pronounced, then a shearing stress must be applied of sufficient magnitude before any flow will occur as, for example, with beaten egg-white or putty. Such a material would be termed *pseudoplastic with yield value* and is represented by curve D in Figure 5.5 where the shearing stress β is the yield value. It is not a liquid in the conventional sense — it behaves as a solid for shear stresses below its yield value and as a pseudoplastic liquid for shear stresses above. It may be termed a *gel*. Generally, if a gel is sheared by a stress greater than its yield value, it will not regain its original yield value on removal of the shear stress. Similarly there are materials that

173

are *dilatant with a yield value*, as represented by curve F in Figure 5.5. Generally, materials with yield values behave as solids in the sense that they can retain their shape indefinitely when subjected to stresses less than their yield value.

Once the yield value has been exceeded, then the material may behave as a Newtonian liquid, as in curve G of Figure 5.5. A material behaving in this manner is termed a *plastic* or a *Bingham body*. Its concept is of great theoretical use as there are many materials that approximate to its ideal behaviour and can be treated without too much mathematical complexity. If μ is the Bingham viscosity where

$$\mu = \frac{\tau - \beta}{S} \tag{5.22}$$

then the apparent viscosity η_{app} is given by

$$\eta_{app} = \frac{\tau}{S} = \mu + \frac{\beta}{S} \tag{5.23}$$

where β is the yield value.

The fact that a material may have a yield value presents many problems in the measurement of viscosity. For example, it means for a material extruded through a tube there will be a central core that moves as a solid plug, or in a rotating cylinder viscometer there will be a layer of stationary material for which the yield value has not been exceeded whilst the rest of the material is behaving as a flowing liquid. Great care must therefore be taken in the interpretation of experimental measurements.

As well as the characteristics of the flow as exhibited in Figure 5.5, many of these liquids show time-dependent flow effects. Some liquids become more fluid with increasing time of flow and are then said to be *thixotropic*. For example, if the liquid is subjected to a constant rate of shear, the decrease in shearing stress to produce this constant rate of shear falls off logarithmically and the apparent viscosity will fall to a limiting low value in a matter of seconds. If the liquid is then left, the apparent viscosity will return to its original value, taking perhaps some hours to do so. An example of this would be salad cream. It is virtually a gel at first but, after the bottle has been shaken to shear the liquid, its viscosity drops to a sufficiently low value to allow it to be poured as a liquid. Conversely, a liquid that shows the reverse properties, that is, a greater resistance to flow with increasing time whilst subjected to a constant rate of shear, is said to exhibit *inverse-thixotropy* or *rheopexy* and the liquid to be inverse-thixotropic or rheopexic.

174

As a general rule it is found that all pure single-phase liquids behave in a Newtonian manner. So also do single-phase liquids containing small or medium-sized molecules and ions, such as a solution of salt and water. If the size of the solute molecule is increased to above about 10^3 atoms, then non-Newtonian properties begin to show themselves. Solutions and melts containing large molecules are always pseudoplastics but not necessarily with a yield value. If the molecules are of long chain polymers, there may be a yield value owing to the need for a shear stress of sufficient magnitude to disentangle and uncoil the long polymer chains. A true and distinct yield value is associated only with an emulsion or a liquid containing a suspension of particles or bubbles, never with a single-phase liquid. The yield arises from the interference between the separate particles. For example, a clay slurry contains individual clay platelets which carry electrostatic charges, with the edge of each platelet being positively, and the face negatively, charged. Thus an open structure of clay platelets interspersed with water forms under static conditions. Before the slurry can flow as a conventional liquid, this structure has to be broken down and the platelets randomly dispersed. To do this requires a certain amount of shearing stress — the yield value. If no such structure is formed, then there will be no yield value. Thus, for example, a slurry that shows considerable settling on being left to stand will not show a yield value.

Dilatant liquids, it is believed, usually consist of suspensions of highly repellent electrostatically charged particles. Owing to this strong mutual repulsion, which keeps the particles in suspension but prevents them forming a bonded structure, there is no yield value unless there is a large volume concentration of particles. A dilatant liquid may be formed when a slurry having the properties of a Bingham body has a deflocculating agent added. This works on the principle of depositing a layer on each particle which gives it a uniform charge, all of the same sign. A dilatant liquid may also be formed if there is only just sufficient liquid to fill the spaces between the particles, which in this case may be uncharged. On shearing, the particles will separate and there may then be insufficient liquid to fill all the spaces and lubricate the particles.

5.6 Applications of Non-Newtonian Fluids

The uses of non-Newtonian fluids are probably more widespread than those of Newtonian fluids, so it is only possible to mention comparatively few representative examples. In foodstuffs, the behaviour of jelly and chocolate are typical of the problems met. Both of these must be firm at room temperature and able to maintain their moulded shapes, but they must also become soft and flow as a liquid when warmed in the mouth. Also of importance is the yield value of butter and margarine. They are required to stay firm over a wide range of temperature but must flow easily to spread.

When oil paints are used for decorating a surface, the paint must be easily spread with a brush and must flow readily to smooth out the brush marks, but it must also become very viscous within a few seconds so that it does not continue to flow under gravity and form 'runs'. Thus paint must have a yield value, with the ideal situation being reached if it also has thixotropic properties. Again, the ink used in ball-point pens must have a yield value and must be insensitive to changes in temperature. It must not flow unless the pen is being used to write otherwise it may drip and stain clothing, nor must it flow too readily when writing otherwise unsightly blobs are left as the writing direction changes. Similar characteristics are required of printing ink, which must spread uniformly on rollers and onto the type. It must then be capable of being transferred uniformly onto paper without smudging. The manufacture of china and pottery requires a clay mixture which is readily moulded into the required shape but will not sag or flow under its own weight prior to firing.

Great importance is paid by the oil industry to the properties of the mud used in drilling operations. As the well is drilled, mud is pumped down the centre of the drill tube where it forces its way round the cutting edges of the drill and back up the annular space between the drill tube and the well-wall. This has a multifold purpose. It cleans away the rock chips and general drilling debris and removes them from the well; it also cools and lubricates the drill bit. In addition it has the important use of preventing caving-in of the well-wall by exerting pressure on the wall. To cause the mud to exert sufficient pressure for this purpose it is usually necessary to raise the density of the mud to about $2500-3000$ kg m^{-3} by adding, for example, iron oxide. As the drilling debris gets mixed into the mud, there is a tendency for its yield value to rise and therefore chemical additives are used to maintain a suitable value by affecting the interactions of the particles, especially the clays. The main purpose of the mud, however, is to remove the drilling debris from the well. Here the yield value is of great importance because the settling out of the debris depends on Stokes' law (equation 4.42). A yield value of the mud is therefore necessary otherwise the debris will start to settle out if the drilling is stopped for any reason. Providing the shear stress due to the weight of the debris is less than the yield value, the debris will remain in suspension. Too high a yield value, however, will also give difficulties — for example, an increased difficulty in pumping, an increased difficulty in forcing the mud through the screens used to remove the debris, as well as difficulties in removing the drill pipe due to suction.

5.7 Lubrication

The need for lubrication has been appreciated from a time shortly after the invention of the wheel and axle. Man at that time must soon have discovered that the application of some animal fat caused easier and quieter running; its use for this purpose has been known for at least 3000 years. However, it has not been until modern times with the use of fast-moving mechanical parts that there has been much under-standing of the problems of lubrication or of the development of more suitable lubricants. Before discussing this further, it is not out of place to look more deeply into the reason for the need of lubricants — friction, and the mechanism of lubrication.

Friction will occur between a solid and a gas, a liquid, or another solid. The first two we have already met in the form of viscosity but it is the friction between two solid bodies that concerns us here. This type of friction can be divided into three categories. The first can be termed *dry* or *solid friction*. This is the friction that exists between two clean dry solids from which all traces of oil or grease, or other contamination, have been removed. A *static coefficient of friction* μ may then be defined as

$$\mu = \frac{F}{R} \qquad (5.24)$$

where F is the maximum frictional force acting as one body is just about to start sliding over the other, and R is the normal force, the force with which the two bodies are being pressed together. In Figure 5.6 a body of weight mg is on the point of sliding down an inclined plane. The force acting down the plane and tending to initiate sliding is $mg \sin \theta$. This is resisted by a force due to friction and acting in a direction to oppose motion of $F = \mu R = \mu \, mg \cos \theta$, that is,

$$mg \sin \theta = \mu \, mg \cos \theta$$

and

$$\mu = \tan \theta \qquad (5.25)$$

177

where θ is the angle of tilt at which slip is just about to occur and μ is the coefficient of static friction between the materials of the block and the plane. For a steel block on a steel plane, with the rubbing surfaces of good machined finishes, this angle θ is about $15°-20°$ and the

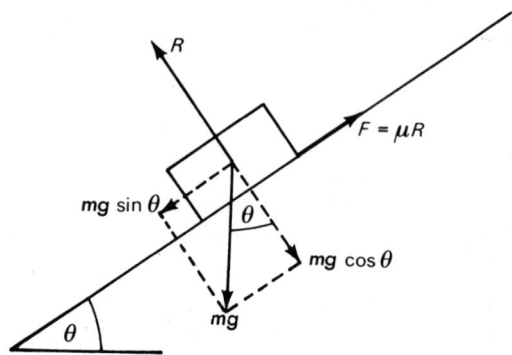

Figure 5.6. *Determination of the coefficient of friction μ as the solid block is about to slide down an inclined plane*

coefficient of friction therefore about $0.27-0.36$. If the angle of tilt is increased, sliding will take place as a series of slips and stops from which little sensible agreement with any theory or repeatability can be obtained.

The second of the categories of friction may be termed *thin-film*, or *boundary*, *friction*. In this case the carefully machined surfaces are separated by a thin film of oil. By thin we usually mean a film perhaps only several molecules thick. With the most carefully machined metal surfaces, however, there are still high spots that protrude through the oil film, so there is an area of metal-to-metal contact of possibly a thousandth part of the opposing interfacial area. As the two components slide over each other, these high spots rub together and local high-temperature spots occur. The shearing stresses involved as two high spots rub together may cause the elastic limit of the metal to be exceeded. Failure will occur at these local spots, with metal being sheared from the peaks. These particles of metal will in turn cause abrasion of other areas. Also, the clean metal surface revealed as the peak of a high spot is removed will come into contact with the clean metal of its opposite number and, with the high local temperature produced, welding will occur at these points. The welds are rapidly fractured again as the sliding bodies move on. Such action thus causes

178

resistance to motion and an overall increase of friction, as well as leading to local high-temperature regions in which the large oil molecules may be cracked, leading to a deterioration in the lubricating qualities of the oil and the production of sludge. Thus it is seen that, in general, boundary film lubrication should be avoided for moving mechanical parts, although some bearings using thin-film lubrication are in successful operation. In these bearings the film is maintained at a sufficient thickness by pressure to avoid metal-to-metal contacts. When the shaft is stationary, its weight tends to squeeze out the oil from underneath so, on starting up again, it may be necessary to force oil into this region to lift the shaft into its central running position surrounded by a uniform thickness of oil.

The third of the categories, and the most important in practice, is that of *fluid-film friction*. Here the moving surfaces are always completely separated by an oil layer, that is, there are no metal-to-metal contacts possible. Figure 5.7 illustrates the difference between boundary

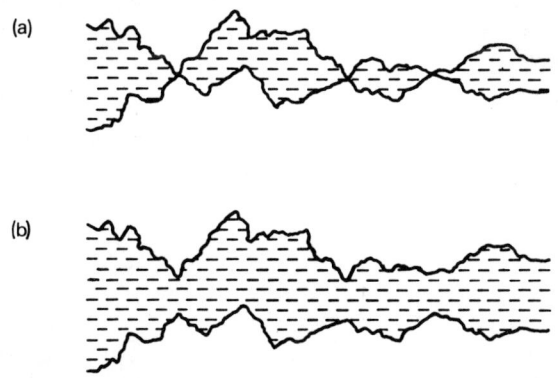

(a)

(b)

Figure 5.7. Illustrating the difference between (a) boundary and (b) thick-film lubrication conditions

and thick-film lubrication conditions. As long as there are stationary layers of liquid in contact with the two surfaces, the only friction between their relative movement is due to the viscosity of the oil. Equation 4.1, Newton's law of viscosity, may then be written in the form to cover this situation as

$$F = \eta\, A\, \frac{v}{d} \qquad (5.26)$$

where F is the force required to maintain uniform motion of one body

179

relative to the other, η the coefficient of viscosity, A the interfacial area, v the relative velocity of motion of the two bodies, and d the thickness of the oil film.

For fluid-film lubrication to be maintained it is necessary for the oil to support the weight of the moving body. Provided the load to be supported is small, a parallel oil film will suffice. The principle of a load-bearing parallel oil film may be seen from the following. In Figure 5.8(a) an oil-covered surface is moving past a stationary fixed

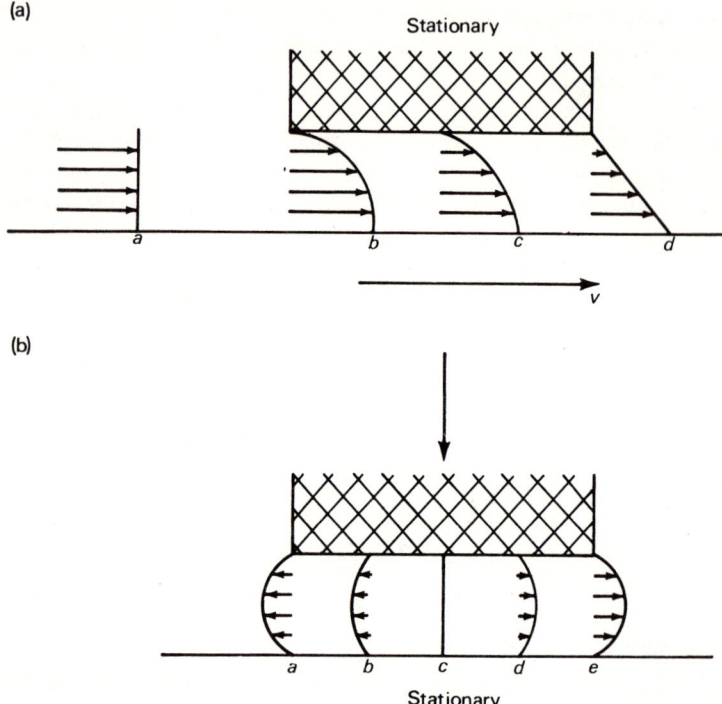

Figure 5.8.   Formation of a load-bearing parallel oil film

body of limited dimensions. A vertical velocity profile at a will be deformed as the oil is dragged past it, as is shown by subsequent profiles at b, c, and d. The change in profile results from the inertial effect of the sudden drop in velocity of the oil layer that meets the stationary body. Owing to the limited size of this stationary body, a uniform velocity gradient cannot be achieved until the velocity profile

180

reaches the exit end to give a linear profile at *d*. Because of this inertial effect on the velocity profile, rather more oil enters the gap than leaves it at the exit end. The excess oil is forced to move at right angles to the direction of motion of the moving body and is thus squeezed out from the sides. The result is the creation of a pressure within the film, which depends on the viscosity and density of the oil and the velocity of the moving body. This pressure enables the oil to support a light load. Such a film is also capable of supporting a load so as to maintain a separation between the bodies even when the motion of the moving body in Figure 5.8(a) is stopped. The situation is then as in Figure 5.8(b). If now the top body starts to move downwards, the oil is squeezed out from the edges with increasing velocity of flow away from the centre. The velocity profiles are then as shown at *a, b, c, d,* and *e*. There is thus a pressure created in the film with a maximum pressure at the centre corresponding to the profile *c* for which there is no outward flow. Thus again a pressure is created which is capable of supporting light loads and preventing the two surfaces coming completely together.

To support greater loads and maintain fluid-film lubrication under more stringent conditions, recourse may be made to the more effective method of the converging oil-wedge. This may be thought of as a combination of the principles depicted by Figures 5.8(a) and 5.8(b). In Figure 5.9(a) we again consider an oil-covered surface moving past a stationary body with velocity profiles as at *a, b,* and *c*. Now considerably more oil enters the interfacial gap than leaves it the other end. This forcing in of oil at the entry end causes a large internal pressure in the interfacial region, with the surplus oil being squeezed out from the sides. There is, however, another effect taking place. As the oil passes through the decreasing gap, it behaves as if the two bodies were moving together, so, by analogy with Figure 5.8(b), oil is squeezed out as shown in Figure 5.9(b). Velocity profiles will then be as shown at *a, b,* and *c,* where the profile *b* nearer to the thin end of the wedge is the line along which there is no outward flow due to squeezing action and consequently is the position of greatest pressure. We may then combine the two effects, as shown in Figure 5.9(c), to give the velocity profiles as shown. By a suitable choice of wedge angle and interfacial area, as well as of viscosity and density of the oil, this wedge film is capable of supporting heavy loads, and its principle is used in many types of thrust and journal bearings and gear mechanisms.

The requirements of the lubricant being known, much may then be done to design suitable ones for specific purposes. The bulk of the

(a)

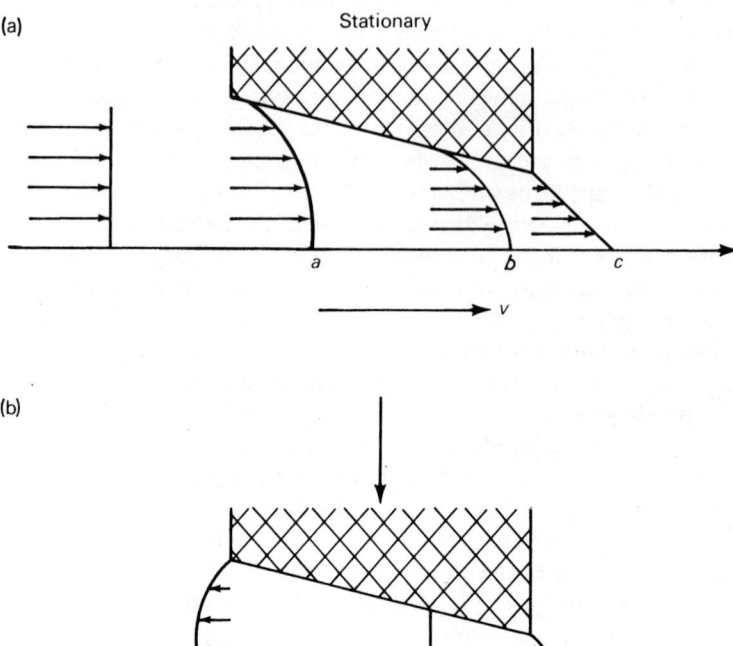

Stationary

a   b   c

v

(b)

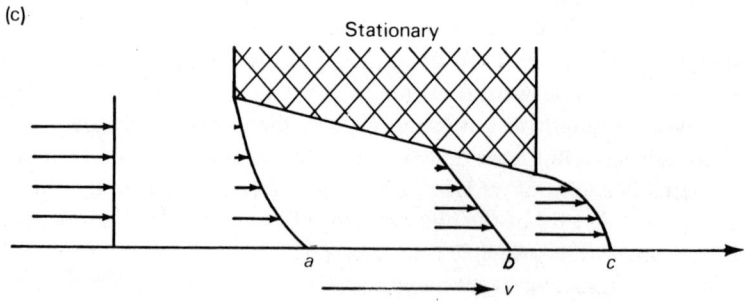

a   b   c

Stationary

(c)

Stationary

a   b   c

v

Figure 5.9.   Action of a wedge-shaped oil film in supporting a load

lubricants used for industrial purposes are hydrocarbon, or mineral, oils. It is essential for economic reasons that an oil should have a long life and in this respect a hydrocarbon oil is very satisfactory. It is chemically stable under a very wide range of operating and storage conditions and unlike the fatty oils, which have animal or vegetable origins, they do not readily combine with oxygen. The fatty oils cannot be distilled, at least at atmospheric pressures, without decomposing. They oxidise on exposure to air and tend to become sticky, with an increase in viscosity, and may become acidic and thus corrosive.

With a mineral oil, however, the problems are much reduced. The more readily oxidisable component can be removed by distillation so that only a relatively small amount of oxidation and decomposition takes place under the extreme working conditions that exist in, for example, an internal combustion engine. Many chemicals may be added to improve their chemical stability without affecting their lubricating qualities. For example, since oxidation of the oil under extreme conditions cannot be completely prevented, inhibitors are added. *Corrosion inhibitors* form protective sulphide and phosphide films on metal surfaces and reduce the corrosion due to acidity of the oil produced by oxidation. Similarly *rust inhibitors* are added to prevent rust forming on cylinder walls due to the deposition of moisture. *Extreme pressure*, or *EP, additives* are used with mineral oils to increase their load-carrying capacity or film strength. These were developed for use with heavily loaded gears and prevent the oil being squeezed out at high-pressure points on the gear teeth. *Detergent additives* in the form of organo-metallic compounds are to prevent the build-up of pitch-like deposits in engines. They are able to dislodge these deposits and keep them dispersed in the oil. It should be remembered that a good lubricating oil does not wear out but becomes contaminated by metallic wear particles and other abrasive or insoluble matter such as dust. Oil is discarded when it contains perhaps one per cent of solid impurities, which could be removed by centrifuging and filtration processes. Water may also accumulate in the oil when it is used, for example, for the lubrication of steam turbines. This water is also conventionally removed by centrifuging.

Obviously one of the most important properties of an oil is its viscosity and, especially, how this varies with temperature. Unfortunately an oil has high viscosity when cold, when it is required to start up an engine, and has a low viscosity when the engine is running and hot. Ideally the viscosity would remain constant with temperature or even perhaps be less viscous at lower temperatures to allow easier starting.

However, this cannot be realised in practice although the temperature dependence on viscosity can be considerably reduced by suitably blending oils and by the use of additives, as in the *multigrade oils.*

So far we have spoken of oils in relation to fluid-film lubrication. If now the lubrication is of the boundary type owing, for example, to high loading or low speeds, then the 'oiliness' of the lubricant becomes of greater importance. Then a great improvement can be obtained by adding colloidal graphite to the oil. It remains dispersed in the oil and is adsorbed on the metal surfaces where it acts as a good 'wetting' agent for the oil and thus assists oil spread to prevent local dry spots. Also, it is a good lubricant in its own right, so where the supply of oil is cut off locally the chance of metal-to-metal contact and welding is considerably reduced. It is also able to resist the high temperatures that may occur in rubbing spots, whereas the oil may be·decomposed at such temperatures. Other colloidal additives operating with the same effects are some molybdenum and tungsten compounds.

As well as mineral oil for lubrication, synthetic oils, in particular the silicones, are used for special purposes. They have the advantage that they are chemically inert and capable of withstanding high temperatures without decomposition. It is possible to manufacture these oils with a great range of viscosities but their cost prohibits their use except for specialist purposes. They are also not suitable for use in conditions where high rubbing speeds or high pressures prevail but otherwise they are quite suitable under fluid-film conditions.

Finally, we may mention greases as lubricants. Greases are a mixture of mineral oils and soaps, with the possible addition of the usual inhibitors and additives. A soap is simply the result of a chemical reaction, termed saponification, between a fat, or fatty acid, and an alkali. The behaviour of the grease depends mainly on the particular alkali metal used in the saponification process and the resulting properties of the soap base. The grease then behaves as a lubricating oil with a yield value. That is, at low rates of shear it behaves as a solid and does not flow. It is thus self-sealing, unlike an oil which requires some mechanical seal. At higher rates of shear it behaves as a normal lubricating oil.

184

Chapter Six

VACUUM PHYSICS

6.1 Introduction

So far we have been concerned mainly with the mechanics of solids and fluids at atmospheric pressures and in fact pressure has had little relevance. If now we discuss gases, then this is no longer the case. In particular, if we consider the behaviour of gases and vapours at low pressures, then we move into the realms of the industrially and practically very important field of vacuum physics — the subject of this chapter.

First of all, why is a vacuum of such interest? To answer this fully would embrace examples from every branch of science and industry. Thus we must satisfy ourselves with some representative examples. If a metal is heated in a vacuum, usually to its molten state but in some cases to just below its melting temperature, it evaporates. Atoms of metal come off as a vapour and travel, owing to thermal energies, away from their hot metal source. They will travel rectilinearly until they collide with gas molecules or with solid objects. With a typical vacuum system and the pressures involved in evaporation processes, the mean free path length of the metal vapour atoms — that is, the average distance travelled before a collision occurs with a gas molecule — is much greater than the dimensions of the vacuum system, so collisions with gas molecules, although some must occur because of random processes, can be ignored. The metal atoms can thus be regarded as travelling rectilinearly from their hot metal source until they hit a solid object. This solid object may well be the chamber walls but can also be any other object introduced into the chamber. Providing the object is not hot, which would cause the metal atoms to be re-evaporated, the object will be coated with evaporated metal atoms which attach themselves one by one. In other words, any object or surface that 'sees' the hot metal source will be coated with that metal. Thus, for example, plastic car trim and plastic jewellery can be 'metallised'. It should be emphasised that an object of any material can be coated in this way. The most common metals used for decorative processes are probably aluminium, gold, copper, and rhodium. Aluminium is commonly used for coating front-reflecting mirrors for telescopes and other optical purposes.

Not only metals can be evaporated but a vast range of other substances as well, for example dielectric materials. Thus, quartz is evaporated onto front-reflecting aluminised mirrors to protect the aluminium layer. Magnesium fluoride is evaporated onto lens surfaces to produce a non-reflecting interference layer, termed 'blooming'. As well as decorative and protective coatings, vacuum evaporation is used extensively in the electronics industry. For example, by depositing a layer of aluminium on an insulating base, followed by a layer of a dielectric material such as quartz or even polyethylene, followed by another layer of aluminium, a thin-film capacitor is made. Similarly a thin layer of aluminium can be used as a thin-film resistor. By evaporating nickel and chromium at the correct rates from adjacent sources, even nichrome thin-film resistors can be made. By depositing a metal layer onto a semiconductor and heat treating to give a controlled diffusion of the metal into the semiconductor, transistor devices are now manufactured. Because the atoms evaporated from a hot source travel in straight lines, an area to be coated can be sharply defined by a mask. So by using successive masking techniques and successive evaporation sources, coupled with chemical etching techniques using photoresists, and by depositing the metal and dielectric layers onto a chip of semiconducting material, complete microcircuits can be made consisting of transistors, capacitors, and resistors. Not to be forgotten also is the use of vacuum techniques in the electronics industry to produce the low pressures required in television tubes and valves.

Increasing use of vacuum processes is being made in the food processing industry. For example, freeze drying is an important method of preservation; here the material is first frozen to retain its structure and then dried under vacuum with the ice passing direct to vapour. Other processes include whisking air bubbles into molten chocolate and suddenly decreasing the surrounding pressure so that these bubbles increase in size. On setting, a chocolate bar is produced with an aerated sponge-like structure which has an interesting texture as well as being large in overall volume and relatively economic in chocolate.

In the chemical industry, use is made of the fact that liquids under vacuum boil at lower temperatures, enabling substances of high molecular weight that would decompose or oxidise at higher temperatures to be distilled. In metallurgy, molten metals may be degassed under vacuum so that on cooling a homogeneous solid without occlusions at the grain boundaries is formed.

In the field of research, vacuum is an important universal tool, whether it be a necessity in the working of an electron microscope or a means of studying physical processes. A sufficiently low vacuum allows a clean uncontaminated surface to be maintained for a convenient period. Thus oxidation processes may be slowed down from virtually instantaneous changes occurring at atmospheric pressure to the same changes taking several hours under ultrahigh vacuum conditions.

The foregoing can serve as only a very brief survey of the uses of vacuum technology but it will be appreciated that uses are widespread and increasing every day. In this chapter we shall introduce the subject with a discussion of aspects of the kinetic theory of gases and go on to discuss the practical aspects of the measurement and production of a vacuum. Before this, however, we must say a little concerning units and ranges of pressure.

In the past vacuum pressure has been measured in terms of several units. The first was in terms of millimetres of mercury (mmHg). This is defined as the pressure exerted by a millimetre-high column of mercury of density $13\,595 \cdot 1$ kg m^{-3} in a place where the intensity of gravity is $9 \cdot 806\,65$ m s^{-2}. That is, 1 mmHg is equivalent to a pressure of $13 \cdot 5951 \times 9 \cdot 806\,65 = 133 \cdot 322$ N m^{-2}. This unit was superseded by the torr, where one torr is 1/760 of the standard atmospheric pressure (atm), defined as $101\,325$ N m^{-2}. That is, 1 torr is equivalent to $101\,325/760 = 133 \cdot 322$ N m^{-2}. Thus for all practical purposes the mmHg and the torr are identical units; in fact they differ by less than

Table 6.1. APPROXIMATE VACUUM PRESSURE RANGES

Vacuum	Pressure range	
	(N m^{-2})	(torr)
Coarse	$10^5 - 10^3$	approx. $10^3 - 10^1$
Medium high	$10^3 - 10^{-1}$	approx. $10^1 - 10^{-3}$
High	$10^{-1} - 10^{-5}$	approx. $10^{-3} - 10^{-7}$
Ultrahigh	$10^{-5} - 10^{-9}$	approx. $10^{-7} - 10^{-11}$

2×10^{-7} torr. Also used in this connection was the micron, equal to 10^{-3} torr. Great Britain, in common with most of Europe, has now adopted SI units and consequently pressures are measured in units of newtons per square metre ($N\,m^{-2}$), where $1\,N\,m^{-2}$ is equivalent to $7 \cdot 500\,64 \times 10^{-3}$ torr (or mmHg). It has been found convenient to divide vacuum pressures into four approximate ranges, owing to the fact that different techniques are required for each of these ranges. These are given in Table 6.1.

6.2 Kinetic Theory and Pressure of Gases

In the previous chapter (Section 5.2) we showed that expressions for the viscosity and thermal conductivity of a gas could be derived from a discussion of the movements of the molecules arising from their thermal energy. In so doing we introduced the concept of the mean free path length (equations 5.1 and 5.2) — the average distance travelled by a gas molecule before it collides with another gas molecule. Similarly, from a consideration of the individual molecules, we can arrive at a better understanding of what is meant by pressure.

Let us then extend the discussion of the kinetic theory of gases given in Section 5.2. The pressure exerted on the chamber walls by a gas may be attributed entirely to the constant bombardment of the walls by the individual molecules of gas. Since the molecules are moving in completely random directions, the individual velocities can be resolved into directions parallel to a system of rectangular axes, x, y, and z, which for simplicity will be regarded as being also parallel to the walls of the chamber. Let an individual molecule have a velocity c in some direction, then this molecule will have velocity components v_x, v_y, and v_z in the x, y, and z directions such that

$$c^2 = v_x{}^2 + v_y{}^2 + v_z{}^2 \qquad (6.1)$$

If this molecule now strikes the wall of the chamber, that is, say, parallel to the yz plane as in Figure 6.1(a) and the collision is assumed to be completely elastic, the component of velocity in the x direction will be reversed. If also the mass of the molecule is m, the change in its x component of momentum will be $2mv_x$ since it changed from mv_x before collision to $-mv_x$ after collision. This process is of course taking place in the chamber with a large number of molecules at any instant. Thus the total change of momentum, that is, the sum of the individual impulses (defined as the total integral of force with respect to time) that occur in unit time and on unit area of the wall, is the pressure exerted by the gas.

188

Let us consider now the number of molecules in a unit volume that have this same component of velocity v_x in the x direction. Let this number be n_i. On average, a half of these will be moving towards the wall and the other half away. If we consider the molecules within a cylinder of volume $v_x \delta t \, \delta A$, as in Figure 6.1(b), where δA is an

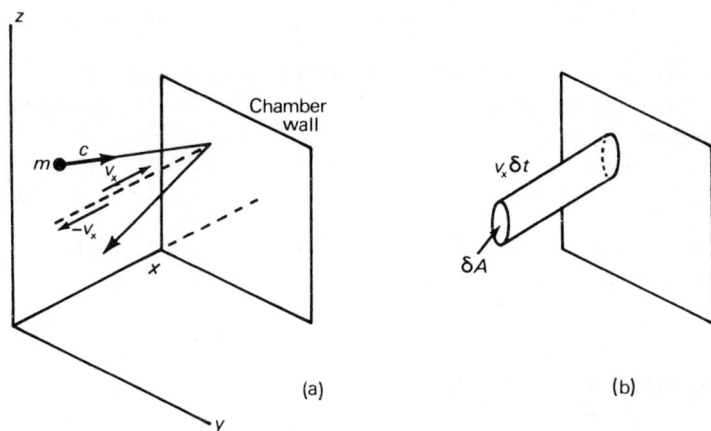

Figure 6.1.   Effect on velocity of a molecule of collision with the chamber wall

element of area of the wall, it can be seen that the molecules contained in this volume and moving towards the wall, that is, $\frac{1}{2} n_i v_x \delta t \, \delta A$, will strike the wall within the time δt. Therefore, since with pressure we are concerned with unit areas, we can say that the number of molecules with a velocity component v_x that strikes unit area of the wall in a time δt is $\frac{1}{2} n_i v_x \, \delta t$, and in unit time $\frac{1}{2} n_i v_x$. We have already shown that each of these collisions results in a momentum change of $2mv_x$, so the total impulse on unit area in unit time is

$$\frac{1}{2} n_i v_x \, (2mv_x) = m n_i v_x{}^2 \tag{6.2}$$

But n_i is the number of molecules per unit volume with a particular velocity component v_x, so, if n is the total number of molecules in unit volume.

$$n = \Sigma \, n_i \tag{6.3}$$

where the summation is over all molecules, each with a corresponding value of v_x. Thus the mean value of $v_x{}^2$ is

$$\overline{v_x{}^2} = \frac{\Sigma \, (n_i v_x{}^2)}{n} \tag{6.4}$$

189

where n_i is the number of molecules with a component of velocity v_x and the summation is over all values of v_x. Hence the pressure, that is, the sum of all the impulses on unit area in unit time due to the molecules with a component of velocity in the x direction, is (from equations 6.2 and 6.4)

$$p = mn\overline{v_x}^2 \qquad (6.5)$$

However, since the directions of motion of the molecules of gas are quite random, the mean values of the components of velocity will be equal, so

$$\overline{c^2} = \overline{v_x}^2 = \overline{v_y}^2 = \overline{v_z}^2 \qquad (6.6)$$

from equation 6.1, where $\overline{c^2}$ is the mean squared velocity of the molecules, and

$$\overline{v_x}^2 = \overline{v_y}^2 = \overline{v_z}^2 \qquad (6.7)$$

Therefore

$$\overline{v_x}^2 = \frac{1}{3}\overline{c^2} \qquad (6.8)$$

Substituting in equation 6.5, the pressure is

$$p = \frac{1}{3}mn\overline{c^2} \qquad (6.9)$$

$$= \frac{1}{3}\rho\overline{c^2} \qquad (6.10)$$

where $\rho = mn$ is the density of the gas at the pressure p. Equation 6.9 can also be written in the form

$$p = \frac{2}{3}n\left(\frac{1}{2}m\overline{c^2}\right) \qquad (6.11)$$

that is, the pressure is equal to two-thirds of the total kinetic energy of the molecules in unit volume. Again, a different form of equation 6.9 is

$$pV = \frac{1}{3}(nV)m\overline{c^2}$$

$$= \frac{1}{3}Nm\overline{c^2} \qquad (6.12)$$

where $N = nV$ is the number of molecules in a volume V.

190

The gas laws follow quite simply from an application of the kinetic theory of gases. Thus, if we consider a mixture of gases of which the components have molecules of masses $m_1, m_2, \ldots$, the partial pressures produced by each of the types of molecules will be

$$p_1 = \frac{1}{3} n_1 m_1 \overline{c_1}^2 = \frac{2}{3} n_1 \left(\frac{1}{2} m_1 \overline{c_1}^2 \right) \qquad (6.13a)$$

$$p_2 = \frac{1}{3} n_2 m_2 \overline{c_2}^2 = \frac{2}{3} n_2 \left(\frac{1}{2} m_2 \overline{c_2}^2 \right) \qquad (6.13b)$$

$$\ldots = \quad \ldots \quad = \quad \ldots$$

where $p_1, p_2, \ldots$ are the individual pressures that would be produced if each of the gases were separately present. The total pressure is produced by bombardment of the chamber walls by all of these molecules and, since kinetic energies are additive, it follows from equations 6.13 that the total pressure is

$$p = p_1 + p_2 + \ldots \qquad (6.14)$$

providing of course there is no chemical reaction between the gases. This is in fact the *law of partial pressures* as first stated by Dalton. We also know that it is heat energy that gives rise to the kinetic energy of the molecules. Therefore it follows from equation 6.12 that, if the number of molecules N is kept constant whilst the volume is allowed to change and the kinetic energy of the molecules is kept constant by keeping the temperature constant, then

$$pV = \text{const.} \qquad (6.15)$$

That is, the pressure of a fixed mass of gas varies inversely as its volume, which is of course *Boyle's law*.

Boyle's law, however, is only absolutely true for what is termed a perfect or ideal gas. In practice there are small departures from the law depending on the temperature and pressure of the gas, but in the region of a specific temperature for a particular gas, termed its Boyle temperature, and for low pressures, a gas behaves as a perfect gas. The gas laws may be regarded as fully applicable to vacuum conditions. Care must, however, be taken when dealing with vapours, of which water vapour is that most commonly associated with vacuum work. In Figure 6.2 is plotted the pressure—volume relationships for a substance such as water or even what is generally regarded as a gas, carbon dioxide. The different curves are characterised by different temperatures. At a temperature t_1, the p–V curve, termed an *isotherm*, has the form *abcd*. Along the section *ab* an increase in the pressure produces a corresponding decrease in volume and the substance behaves as a gas until the point *b* is reached. Any further attempts to increase the pressure will cause some of the gas to liquefy with a consequent decrease in volume. Thus along the section *bc* the pressure remains

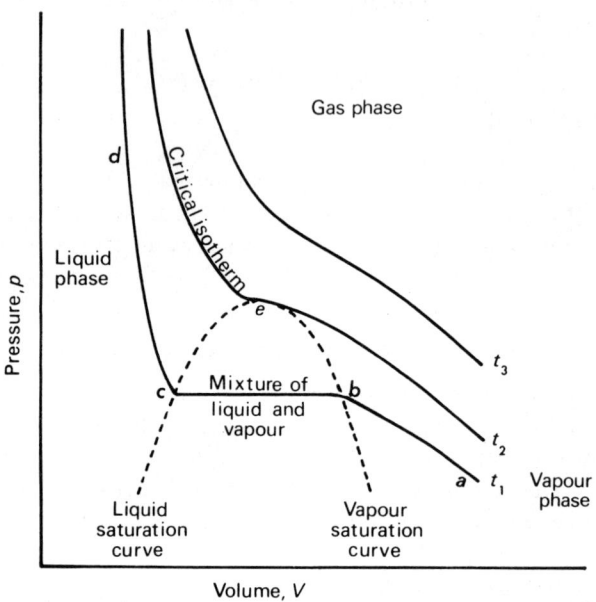

Figure 6.2. Pressure—volume relationship showing behaviour of vapours

constant with the volume decreasing as more and more gas is liquefied to counteract any attempts to increase the pressure, until the condition is reached at point *c* when all the gas is liquefied. The section *cd* thus represents a liquid phase in which large increases in pressure produce only small reductions in volume.

The isotherm corresponding to the temperature t_2 is termed the *critical isotherm*. It is the lowest-temperature isotherm for which no liquefaction occurs no matter how much the pressure is increased. The dotted curve through *b* is the locus of points representing the conditions at which liquefaction of a vapour begins, and similarly the dotted curve through *c* is the locus of points at which evaporation of a liquid begins. These loci meet at the point *e* on the critical isotherm, termed the critical point. It is at this point that the specific volumes of the liquid and vapour become equal. From the figure it is also seen that the substance can pass from an obviously gaseous state to an obviously liquid state without any discontinuity or mixed state occurring. Thus we are able to differentiate between a vapour and a gas. A gas is a material that is above its critical temperature, such as at t_3, so it cannot be liquefied no matter how great the pressure, whereas a vapour can be the same material but this time below its critical temperature. On compressing a vapour sufficiently, liquefaction will occur.

As has already been said, water vapour is commonly associated with vacuum systems. Water has a critical temperature of 374°C, so any water vapour in a vacuum system will liquefy when the pressure is increased sufficiently. Thus any pressure measurements involving the compression of a known volume of gas and the application of Boyle's law may well be nullified if water vapour is present in the volume.

In pumping down a chamber to create a vacuum, it must be appreciated that a vast number of molecules must be removed and, even when an ultrahigh vacuum has been obtained, there still remains a vast number. To obtain some appreciation of the quantities involved, let us now consider a kilogram-mole (kg-mole) of gas, that is, a mass of gas equal to the molecular mass expressed as an equivalent number of kilograms. Then the ideal gas equation, which has been determined empirically, can be written as

$$pV = RT \qquad (6.16)$$

where p is the pressure in newtons per square metre, V is now the volume occupied by 1 kg-mole of gas, R is the gas constant, equal to 8.314×10^3 J kg-mole^{-1} K^{-1}, and T is the absolute temperature of the gas in kelvin. This equation is analogous to equation 6.12 determined from the kinetic theory of gases. Equation 6.12 may be rewritten as

$$pV = \frac{1}{3} N_A m \overline{c^2}$$
$$= \frac{2}{3} N_A \left(\frac{1}{2} m \overline{c^2} \right) \qquad (6.17)$$

where V is also now the volume occupied by 1 kg-mole and N_A is the number of molecules in 1 kg-mole, that is, the number of molecules in a volume V as before. N_A is the same number for all substances and is in fact Avogadro's constant, equal to 6.023×10^{26} molecules kg-mole^{-1}. From equations 6.16 and 6.17 it can be seen that

$$\tfrac{1}{2} m\overline{c^2} = \frac{3R}{2N_A} T$$

$$= \tfrac{3}{2} kT \qquad (6.18)$$

where $k = R/N_A$ is called Boltzmann's constant. Equation 6.18 also gives a meaning to the gas constant R. It shows that R is equal to two-thirds of the total kinetic energy of the gas molecules in 1 kg-mole of gas at 1 K. It also follows from equation 6.18 that

$$\sqrt{(\overline{c^2})} = c_{\text{r.m.s.}} = \sqrt{\frac{3kT}{m}} = \sqrt{\frac{3RT}{N_A m}} \qquad (6.19)$$

where $c_{\text{r.m.s.}}$ is the 'root-mean-square' speed of the molecules, each of mass m. If we consider, say, nitrogen with a molecular mass of 2 x 14 atomic mass units, where one atomic mass unit is equal to $1 \cdot 660 \times 10^{-27}$ kg, and therefore with a molecular mass of 2 x 14 x $1 \cdot 660 \times 10^{-27} = 4 \cdot 65 \times 10^{-26}$ kg, we may calculate the root-mean-square speed of these molecules at, say, $0°C$ (273 K). Then

$$c_{\text{r.m.s.}} = \sqrt{\frac{3 \times 8 \cdot 314 \times 10^3 \times 273}{6 \cdot 023 \times 10^{26} \times 4 \cdot 65 \times 10^{-26}}}$$

$$= 4 \cdot 93 \times 10^2 \text{ m s}^{-1}$$

which is over 1100 miles per hour!

We can also quite easily determine the number of molecules of gas in a unit volume from equation 6.16. Rewriting this equation for a volume measured in cubic metres,

$$pV = \frac{N}{N_A} RT = NkT \qquad (6.20)$$

where p is still the pressure in newtons per square metre, N is the total number of molecules in the volume V measured in cubic metres, and N_A, R, T, and k are Avogadro's number, the gas constant, the absolute temperature, and Boltzmann's constant respectively as before. Hence the number of molecules n in one cubic metre of gas is

$$n = \frac{N_A p}{RT} \qquad (6.21)$$

$$= \frac{6 \cdot 023 \times 10^{26}}{8 \cdot 314 \times 10^3} \frac{p}{T}$$

$$= 7 \cdot 244 \times 10^{22} \frac{p}{T} \text{ molecules}$$

194

Thus, for example, at standard temperature and pressure, that is, at $0°C$ (273 K) and 101 325 $N\,m^{-2}$ (1 atm), the number of molecules per cubic metre is $7.244 \times 10^{22} \times 101\,325/273 = 2.69 \times 10^{25}$. How this number n varies at room temperature, say, $20°C$ (293 K), with the vacuum pressure is shown graphically in Figure 6.3.

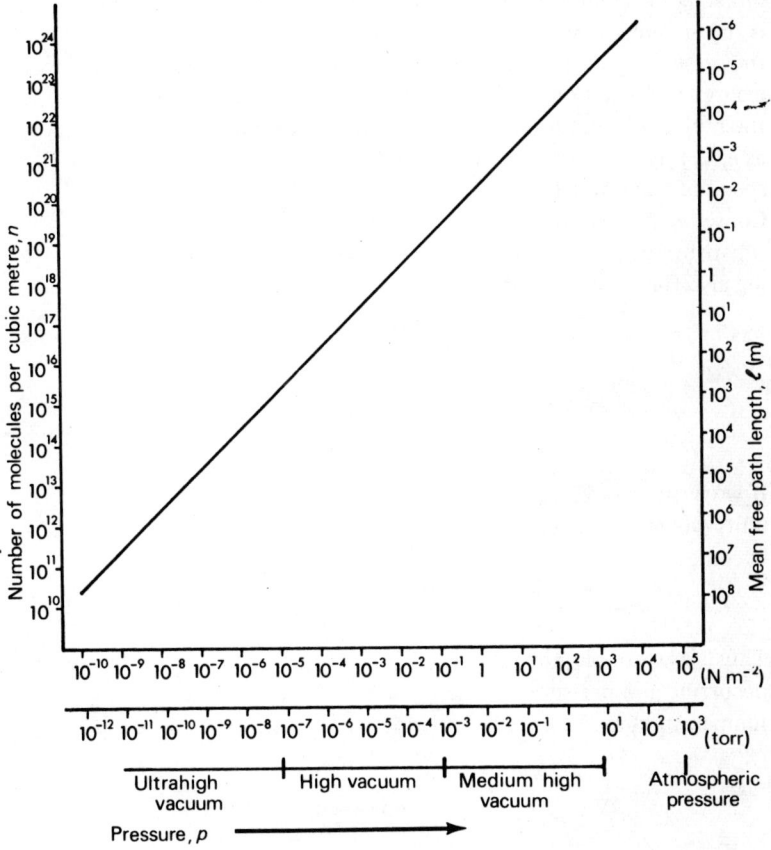

Figure 6.3.   Variation of number of molecules and mean free path length of nitrogen at $20°C$, with pressure

Another calculation of interest, that again gives an idea of the numbers involved, is for the time required to produce a monomolecular layer over unit area. This is of especial interest in oxidation studies. Knudsen has shown from the kinetic theory of gases that the number

195

of gas molecules striking unit area of the chamber wall in unit time is given by

$$n' = \frac{1}{4}n\bar{c} \qquad (6.22)$$

where n is the number of gas molecules in unit volume. It is assumed that the pressure is low so that the mean free path length of the gas molecules is great compared with the chamber dimensions. The term $\bar{c}$ is the arithmetic mean of the speeds of the molecules; it is not quite the same as $c_{\text{r.m.s.}}$ that has occurred in our equations up to now, for example in equation 6.19. Thus $\bar{c}^2$ is not the same as $\overline{c^2}$. However, for the molecular speeds occurring in a gas these are not so very different, as $c_{\text{r.m.s.}}$ is approximately equal to $1 \cdot 09\bar{c}$. That is, the root-mean-square speed of a gas molecule is about 9% greater than the average speed. Consequently we shall not be introducing any appreciable error in substituting $c_{\text{r.m.s.}}$ from equation 6.19 for $\bar{c}$ in equation 6.22. Substituting also for n from equation 6.21, equation 6.22 becomes

$$n' = \frac{N_A p}{4RT} \sqrt{\frac{3RT}{N_A m}}$$

$$= \frac{p}{4} \sqrt{\frac{3N_A}{RTm}} \qquad (6.23)$$

If we write $m = M/N_A$, where M is the molecular weight in atomic mass units whereas m is the molecular weight in kilograms, equation 6.23 becomes

$$n' = \frac{N_A p}{4} \sqrt{\frac{3}{RTM}} \qquad (6.24)$$

Thus, for example, the number of impacts per square metre per second occurring at a pressure of p newtons per square metre at room temperature, say, 20°C (293 K), and with nitrogen for which $M = 28$, is

$$n' = \left(\frac{6 \cdot 023 \times 10^{26}}{4} \sqrt{\frac{3}{8 \cdot 314 \times 10^3 \times 293 \times 28}} \right) p$$

$$= 3 \cdot 2 \times 10^{22} \, p \text{ impacts m}^{-2} \qquad (6.25)$$

However, if we had substituted the correct speed (the average speed) in equation 6.22 instead of its near approximation (the root-mean-square speed), we would have obtained the more correct expression

$$n' = 2 \cdot 9 \times 10^{22} \, p \text{ impacts m}^{-2} \qquad (6.26)$$

instead of equation 6.25. The difference, it is thus seen, is negligible.

196

Let us now return to the original problem of determining the time for a monomolecular layer to be formed on a surface at room temperature. For a gas molecule to attach itself to, say, a metal surface, we may assume that it will do so at an existing atomic site. In other words, the gas layer forms a lattice-like structure that is an extension of the crystalline lattice of the metal. For a typical metal the interatomic distance is of the order $2 \cdot 5 \times 10^{-10}$ m, giving a possible $1 \cdot 6 \times 10^{19}$ atomic sites per square metre. If every gas atom that hit this surface attached itself to an atom site, then it would take a time t to fill every site, that is, to form a monolayer one metre in area, where

$$t = \frac{1 \cdot 6 \times 10^{19}}{2 \cdot 9 \times 10^{22} p}$$

$$= \frac{5 \cdot 5 \times 10^{-4}}{p} \text{ s} \qquad (6.27)$$

For a high vacuum of, say, 10^{-4} N m^{-2}, this time will be about $5\frac{1}{2}$ s increasing to over 15 h at an ultrahigh vacuum pressure of 10^{-8} N m^{-2}. This then is an important reason for the use of ultrahigh vacuum equipment — it enables surfaces to remain clean and uncontaminated for a sufficient time to enable experiments to be carried out.

Not all molecules or surfaces, however, allow every incident molecule to remain attached. Some do not attach themselves in the first place and others detach themselves again after a short time. For metals what is termed a *sticking coefficient* has values mainly between $0 \cdot 1$ and unity. It is a measure of the probability of a gas molecule striking the surface and remaining for an infinite time.

6.3 Conductance and Pumping Speed

To evacuate a chamber by means of a vacuum pump requires a pipe connection between the chamber and the pump. With the chamber starting at atmospheric pressure, the rush of air through the pipe as the pump is switched on creates a turbulent flow pattern through the pipe. This, however, is of little interest because with the falling pressure the turbulent flow very soon ceases and streamline flow conditions prevail. In the case of vacuum physics, this streamline condition is termed *viscous flow*. Then the mean free path length is still short relative to the dimensions of the tube diameter. We already have an expression for the mean free path length, derived in the previous chapter (equation 5.2),

$$l = \frac{1}{n\pi d^2 \sqrt{2}}$$ (6.28)

where n is the number of molecules in a cubic metre and d is the diameter of a gas molecule. From equations 5.2 and 5.6, coupled with a knowledge of viscosities, mean velocities, and molecular masses, the diameters of molecules can be calculated. Some values of gas-molecule diameters are given in Table 6.2.

Table 6.2. DIAMETERS OF SOME GAS MOLECULES, DETERMINED FROM VISCOSITY MEASUREMENTS

Gas	Diameter $(10^{-10}$ m)
Argon	3·4
Carbon dioxide	3·9
Helium	2·6
Hydrogen	3·0
Nitrogen	3·7
Oxygen	3·5

If we accept a representative value of $3\cdot7 \times 10^{-10}$ m for a diameter (that is, nitrogen) and substitute into equation 6.28 this and values of n, the number of molecules in a cubic metre from equation 6.21, we can calculate the mean free path lengths at various pressures. Then the mean free path length at an absolute temperature T and a pressure p is

$$l = \frac{2\cdot27 \times 10^{-5} \, T}{p} \text{ m}$$ (6.29)

If we assume a value for room temperature of, say, $20°C$ (293 K), this expression becomes

$$l = \frac{6\cdot65 \times 10^{-3}}{p} \text{ m}$$ (6.30)

where the mean free path length l is in metres and the pressure p in newtons per square metre. The variation of the mean free path length with pressure for nitrogen (and hence approximately air) at room temperature is shown graphically in Figure 6.3. Thus it can be seen that at a pressure of about $0\cdot3$ N m^{-2} the mean free path length of the

198

gas molecules is about 0·02 m, which is probably a similar dimension to the diameter of the connecting pipes.

At lower pressures still, the mean free path length becomes great compared with the dimensions of the apparatus and the molecules move almost completely independently of one another. Collisions with the walls now become much more frequent than collisions with other molecules. Gas flow in this condition is termed *molecular flow*. At intermediate pressures, when the mean free path length is of the same order as the diameter of the pipe through which the gas flows, the flow is dependent on both molecular and viscous properties and the flow is then said to be in the *transition range*.

It is now useful to make use of the *conductance C*, which can be expressed as

$$\text{conductance, } C = \frac{\text{pressure at any point x volume flow at that point}}{\text{pressure drop}}$$

$$(6.31)$$

where the flow is expressed in cubic metres per second measured at the particular pressure at that point. Hence we can give an expression for what may be termed the 'throughput' Q, where

$$Q = C(p_1 - p_2)$$

$$= \text{pressure x volume flow} \qquad (6.32)$$

where $p_1 - p_2$ is the pressure drop over which the conductance is measured. The units of the throughput will be newton metre per second $(N\,m\,s^{-1})$. For viscous flow, when the mean free path length is small relative to the diameter of the conducting pipe, we can apply the extension to the Hagen—Poiseuille equation for gases (equation 5.17). Then the conductance of a pipe of length l, diameter d, and with a throughput Q, equal to the product of volume flow and pressure, is given by

$$C = \frac{Q}{p_1 - p_2}$$

$$= \frac{\pi(p_1 + p_2)\,d^4}{256\,\eta l}$$

$$= \frac{\pi p d^4}{128\,\eta l} \qquad (6.33)$$

Here we are using the diameter instead of the radius in equation 5.17

199

and the pressure p is the mean pressure along the pipe, that is, $p = (p_1 + p_2)/2$ where p_1 and p_2 are the pressures at the two ends. Equation 6.33 simplifies to

$$C = 2.45 \times 10^{-2} \frac{pd^4}{\eta l} \text{ m}^3 \text{ s}^{-1} \qquad (6.34)$$

With a viscosity of air at $20°C$ (293 K) of $\eta = 1.81 \times 10^{-5}$ N s m^{-2}

$$C = 1.35 \times 10^3 \frac{pd^4}{l} \text{ m}^3 \text{ s}^{-1} \qquad (6.35)$$

for the conductance of a pipe.

For molecular flow, when the mean free path length is great compared to the diameter of the pipe, Knudsen has derived, on a partly empirical basis, values for conductances. These are given in Table 6.3 for some

Table 6.3. FORMULAE FOR CALCULATING CONDUCTANCES*

Conductance	Formula
Viscous conductance Round pipe	$1.35 \times 10^3 \ pd^4/l$
Molecular conductance Round pipe Orifice of any shape	$121 \ d^3/(l + 1.3d)$ $116 \ A$

*Conductances in units of m^3 s^{-1}.
Pressure p in units of N m^{-2}.
Diameter of pipe d in metres.
Length of pipe l in metres.
Orifice area A in square metres.

common situations for gas temperatures of $20°C$ (293 K). It will be seen that molecular conductances are independent of the pressure. For high conductance it is seen from the table that it is essential to employ as large a diameter connecting pipe as possible, coupled with as short a pipe as possible. If the conductance is limited by the pipe, then no matter how great the capacity of the pump, virtually no faster pumping can be achieved.

Conductance formulae are not necessary for the transition range of mixed molecular and viscous flow. It is found to be quite sufficient to use the formulae for viscous flow conductance down to a transition pressure p_t, where

$$p_t = \frac{7 \cdot 5 \times 10^{-2}}{d} \, \text{N} \, \text{m}^{-2} \tag{6.36}$$

and where d is the diameter of the pipe in metres, and then to use molecular flow conductances for lower pressures. The transition range in fact extends from $10p_t$ to $0 \cdot 1 p_t$. For systems in which different flow conditions prevail, it is found that

total conductance = molecular conductance + viscous conductance

$$\tag{6.37}$$

Also, if several conductors are in series, the total conductance C is given by

$$\frac{1}{C} = \frac{1}{C_1} + \frac{1}{C_2} + \frac{1}{C_3} + \ldots \tag{6.38}$$

where $C_1, C_2, C_3, \ldots$ are the individual conductances. Similarly, for conductors in parallel

$$C = C_1 + C_2 + C_3 + \ldots \tag{6.39}$$

Let us now define a pumping speed S_P of a pump. It can be defined by the expression

$$Q = p \, S_P \tag{6.40}$$

where the throughput Q, by comparison with equation 6.32, is equal to the product of the pressure p, in this case measured at the input to the pump, and the volume per unit time of gas being removed by the pump, again measured at the pressure of the pump inlet. In other words, the pumping speed is the volume of gas removed by the pump in unit time, measured at the pressure of the pump inlet. If the pump is separated from the chamber by a pipe or orifice of conductance C, the effective speed of the pump becomes S where

$$\frac{1}{S} = \frac{1}{S_P} + \frac{1}{C} \tag{6.41}$$

As an example we may consider the problem of evacuating a small electronic valve, which owing to its size allows only a small-bore pipe connection to the pump. From equation 6.41, the effective pumping speed at the chamber is

$$S = \frac{S_P C}{S_P + C} \tag{6.42}$$

where S_P is the pumping speed of the pump itself and C is the conductance of the connecting pipe. Supposing a pump is chosen which has a speed 10 times the conductance of the pipe, that is, $S_P = 10C$, then from equation 6.42 the effective speed is

$$S = \frac{10}{11} C$$

If instead we had used a pump with 10 times the speed of the first one, so that $S_P = 100C$, the effective speed would be increased to

$$S = \frac{100}{101} C$$

This is only an increase of about 9% in the effective speed for a tenfold increase in the pump speed, so the extra expense of the larger pump would not be justified.

From Boyle's law, for a small change δp in pressure with its corresponding volume change δV,

$$pV = (p - \delta p)(V + \delta V)$$
$$= pV + p\,\delta V - V\delta p - \delta p\,\delta V \qquad (6.43)$$

Hence

$$p\,\delta V - V\delta p = 0$$

since $\delta p\,\delta V$ is negligibly small. Then

$$\delta p = \frac{p}{V}\,\delta V$$

and

$$\frac{dp}{dt} = \frac{p}{V}\frac{dV}{dt} \qquad (6.44)$$

However, by definition the speed S is given by

$$S = -\frac{dV}{dt} \qquad (6.45)$$

where the negative sign arises since the volume removed decreases as time increases, that is, the speed gets less with time. Then, from equations 6.44 and 6.45, the rate of reduction of pressure is

$$\frac{dp}{dt} = -\frac{S}{V}p \qquad (6.46)$$

202

At the ultimate vacuum pressure of the system, that is, when the pressure reaches an equilibrium state due to the limiting nature of the pump and any leaks, dp/dt becomes zero and therefore so does the pumping speed S from equation 6.46.

We can also extend equation 6.46 in the following way. First, this equation can be written as

$$S \, dt = -\frac{V}{p} \, dp \qquad (6.47)$$

Then, if p_1 and p_2 are pressures measured at times t_1 and t_2, and assuming the pumping speed S remains constant over this pressure range, equation 6.47 can be integrated as

$$S \int_{t_1}^{t_2} dt = -V \int_{p_1}^{p_2} \frac{1}{p} \, dp \qquad (6.48)$$

that is,

$$S(t_2 - t_1) = -V(\log_e p_2 - \log_e p_1)$$

Therefore

$$\begin{aligned} S &= \frac{V}{t_2 - t_1} \log_e \left(\frac{p_1}{p_2}\right) \\ &= \frac{2 \cdot 303 \, V}{t_2 - t_1} \log_{10} \left(\frac{p_1}{p_2}\right) \qquad (6.49) \end{aligned}$$

This equation enables the speed S to be measured conveniently in units of $m^3 \, s^{-1}$, since the pressures and corresponding times are easily measured together with the volume $V \, m^3$, the volume being evacuated. Conversely, knowing the speed over this pressure range, we can calculate the time required to reduce the pressure from p_1 to p_2. It should be remembered that S is the effective speed of the pump at the chamber, taking into account the conductance of the pipe connections as in equation 6.42.

As an example of the use of the above equations, we shall calculate the lowest pressure that can be attained in a chamber that is connected to a pump with a pumping speed of $5 \times 10^{-2} \, m^3 \, s^{-1}$ by a pipe 0.02 m in diameter and 0.2 m long and with a lowest pressure of $1 \times 10^{-4} \, N \, m^{-2}$ attainable at the pump inlet. First of all we must calculate the conductance of the pipe. From equation 6.36, the transition pressure between viscous and molecular conductance is

$$\begin{aligned} p_t &= \frac{7 \cdot 5 \times 10^{-2}}{0 \cdot 02} \\ &= 3 \cdot 75 \, N \, m^{-2} \end{aligned}$$

Obviously we are working well below this pressure and the conductance is therefore (from Table 6.3)

$$C = \frac{121\,d^3}{l + 1 \cdot 3\,d}$$

$$= \frac{121 \times 0 \cdot 02^3}{0 \cdot 2 + (1 \cdot 3 \times 0 \cdot 02)}$$

$$= 4 \cdot 3 \times 10^{-3} \text{ m}^3 \text{ s}^{-1}$$

From equation 6.42 the effective pumping speed at the chamber is

$$S = \frac{S_P C}{S_P + C}$$

$$= \frac{(5 \times 10^{-2})\,(4 \cdot 3 \times 10^{-3})}{(5 \times 10^{-2}) + (4 \cdot 3 \times 10^{-3})}$$

$$= 4 \cdot 0 \times 10^{-3} \text{ m}^3 \text{ s}^{-1}$$

Now we can write, in agreement with Boyle's law,

$$pS = p_P S_P$$

where p is the pressure at a point where the speed, that is, the volume flowing per second, is S; p_P and S_P are the corresponding pressure and speed at the pump inlet. Then the pressure at the chamber is p where

$$p = \frac{p_P S_P}{S}$$

$$= \frac{(1 \times 10^{-4})\,(5 \times 10^{-2})}{4 \cdot 0 \times 10^{-3}}$$

$$= 1 \cdot 3 \times 10^{-3} \text{ N m}^{-2}$$

Alternatively, using equation 6.31,

$$\text{pressure drop} = \frac{\text{pressure at any point} \times \text{volume flow at that point}}{\text{conductance, } C}$$

$$= \frac{(1 \times 10^{-4})\,(5 \times 10^{-2})}{4 \cdot 3 \times 10^{-3}}$$

$$= 1 \cdot 2 \times 10^{-3}$$

$$= 12 \times 10^{-4} \text{ N m}^{-2}$$

This is the pressure drop along the pipe and hence the pressure in the chamber is

$$p = (1 \times 10^{-4}) + (12 \times 10^{-4})$$

$$= 13 \times 10^{-4}$$

$$= 1 \cdot 3 \times 10^{-3} \, \mathrm{N\,m^{-2}}$$

as before.

We can also make calculations of the time involved in pumping down the chamber. Supposing, for example, we wish to know the time to reduce the pressure in the chamber from $10^{-1} \, \mathrm{N\,m^{-2}}$ to $10^{-3} \, \mathrm{N\,m^{-2}}$, assuming the pumping speed remains substantially constant over this pressure range, where the chamber has a volume of, say, $0 \cdot 1 \, \mathrm{m^3}$. Then from equation 6.49 the time taken will be

$$t_2 - t_1 = \frac{2 \cdot 3 \, V}{S} \log_{10}\left(\frac{p_1}{p_2}\right)$$

$$= \frac{2 \cdot 3 \times 0 \cdot 1}{4 \cdot 0 \times 10^{-3}} \log_{10}\left(\frac{10^{-1}}{10^{-3}}\right)$$

$$= 1 \cdot 2 \times 10^2 \, \mathrm{s}$$

$$= 2 \, \mathrm{min}$$

6.4 Vacuum Gauges

Before going on to describe actual pumps and complete pumping systems, it is perhaps most fitting if we first deal with gauges that are used to measure the pressures produced in vacuum systems. Once a pressure is reached below that which may be practically measured by a manometer, there is a lack of absolute gauges. In fact the only gauge for vacuum pressures that can be regarded as completely absolute is the *McLeod gauge.* Its operation is based purely on Boyle's law and it is against this gauge that all other low-pressure gauges are calibrated, often with considerable extrapolation.

Figure 6.4(a) shows an arrangement of the gauge. With it evacuated, a column of mercury some 76 cm high can be supported, and consequently facilities must exist for lowering the mercury reservoir by this amount to clear the mercury from the gauge. An alternative arrangement for doing this, shown in Figure 6.4(b), uses a vacuum pump to

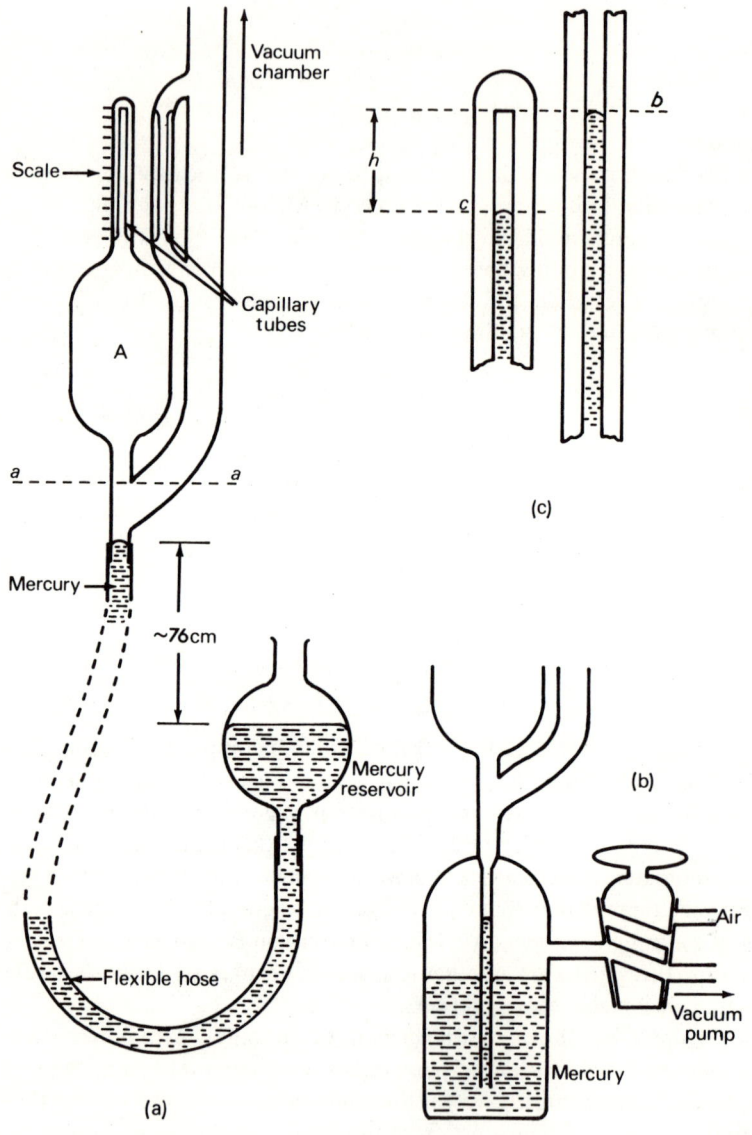

Figure 6.4. Arrangement of a McLeod gauge

reduce this head of mercury. To measure a pressure, the mercury column is raised by raising the mercury reservoir or by allowing air into the mercury flask in the alternative arrangement. When the mercury reaches the level aa, it traps in chamber A a volume V of gas at the vacuum pressure p to be measured. The mercury is allowed to rise to the height b in the comparison tube [Figure 6.4(c)], which is level with the top of the closed capillary tube. The mercury in the closed capillary tube then reaches some level c, where the difference between levels c and b is h metres. Thus the volume V of gas at pressure p has now been compressed to some new volume $V' = Ah$, where A is the cross-sectional area of the capillary tube, with a pressure p' equal to the vacuum pressure p plus the pressure exerted by the mercury column of height h. Therefore, by Boyle's law,

$$pV = p'V'$$

$$= (p + \rho gh) Ah \tag{6.50}$$

where ρ is the density of mercury, $1 \cdot 3595 \times 10^4$ kg m^{-3}, and g the intensity of gravity, $9 \cdot 8067$ m s^{-2}, so $\rho gh = 1 \cdot 3332 \times 10^5 h$ N m^{-2}. Generally, however, the vacuum pressure p is negligible in comparison with the mercury pressure ρgh, and equation 6.50 can be written as

$$p = \frac{\rho gAh^2}{V}$$

$$= 1 \cdot 33 \times 10^5 \frac{Ah^2}{V} \text{ N m}^{-2} \tag{6.51}$$

where A square metres, the cross-sectional area of the measuring capillary, and V cubic metres, the volume of gas initially trapped by the mercury column, are constants determined for the particular gauge. Thus, by measuring the height h in metres, the pressure is readily determined.

This gauge has the obvious advantages of simplicity, particularly in its theory which is independent of the nature of the gas, and of absolute calibration. However, it does have a number of disadvantages. Apart from the fact that pressures cannot be read from it continuously or instantaneously — to read the pressure again, the mercury column must be lowered to below the level aa and then brought up again — it does not give the correct pressure when condensable vapours are present. Thus considerable error can result if water vapour is present in the system. At sufficiently low pressures, mercury vapour itself becomes of importance. At about room temperature, mercury has a vapour pressure of $0 \cdot 160 \ \mathrm{N\,m^{-2}}$, and consequently the vacuum chamber cannot be pumped down to a lower pressure than this unless a cold trap such as that shown in Figure 6.5 is used to isolate the McLeod gauge from the rest of the vacuum system.

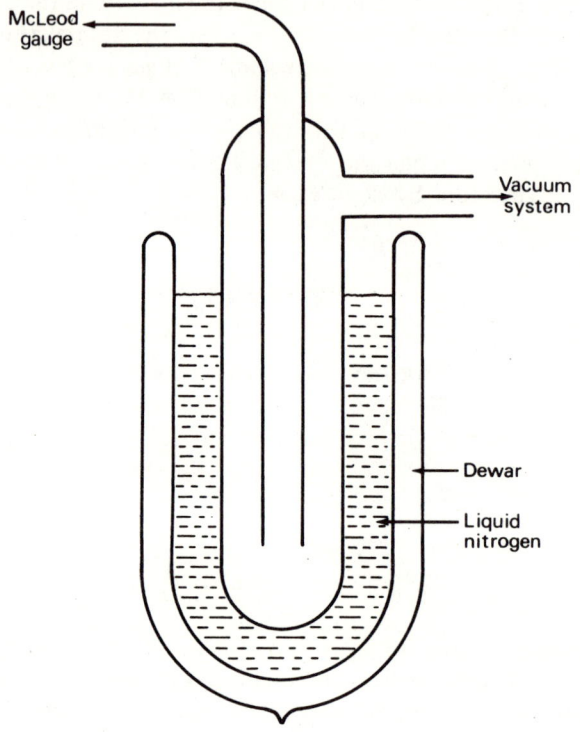

Figure 6.5.   Design of a cold trap to prevent mercury vapour from a McLeod gauge entering the rest of the vacuum system

The lower pressure limit of the gauge can be extended by increasing the volume of the chamber A and by decreasing the cross-sectional area of the capillary. Both of these actions have practical limitations — the first due to inconvenient size and volume of mercury, the second due to sticking of the mercury thread in the capillary. In practice, the McLeod gauge, although an absolute gauge, is not suitable for routine use but is used to calibrate other gauges that are more suitable.

A very useful gauge is the *Pirani gauge*. This takes over from the mercury manometer or, as is more usual, it is used as the first vacuum gauge of the system since in practice the pressure is down to the Pirani range (see Figure 6.6) within a minute or so of switching on the pumps. The gauge element is simply a hot filament which is generally heated only to a dull red to prevent burn-out. Changes in gas pressure

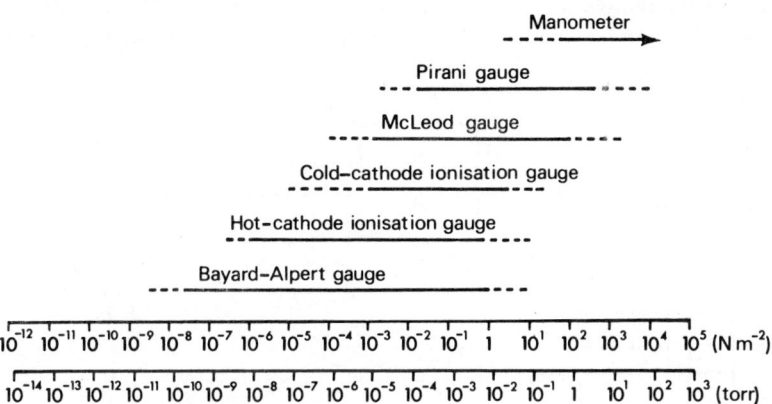

Figure 6.6. Range of pressures measurable with commonly used gauges – the full line indicates the usual range

affect the rate at which heat is conducted away from the filament and consequently affect its temperature. The filament material is chosen to have a high temperature coefficient of resistance so that changes in the gas pressure manifest themselves as a measurable change in resistance. The resistance of the hot filament is measured by a Wheatstone bridge which is balanced with the filament in a vacuum lower than the operating range of the gauge. At higher pressures the balance is upset and the bridge galvanometer then records an out-of-balance current, which is a measure of the gas pressure. The upper pressure limit is due to the thermal conductivity of the gas becoming mainly independent of the pressure at higher values, whereas the lower limit is due to radiation losses becoming relatively more important than conduction by the gas. Since the calibration of the gauge is dependent on the thermal conductivity of the gas at a particular pressure, the calibration is upset when the gauge is used with a gas of different thermal conductivity.

For lower pressures than those measurable with a Pirani gauge, ionisation gauges come into their own. These work on the principle of producing positive ions by the bombardment of neutral gas molecules by electrons. The positive ion current produced is thus a function of the number of collisions, which in turn is a function of the pressure. Two methods are used to produce the bombarding electrons — from a cold cathode or from a hot filament.

The *cold-cathode ionisation gauge*, termed a *Penning*, or *Philips* (after the original manufacturer), *gauge*, comprises a ring anode with a cathode mounted on either side, as shown in Figure 6.7. The anode—cathode assembly is enclosed within a glass or a metal envelope connected

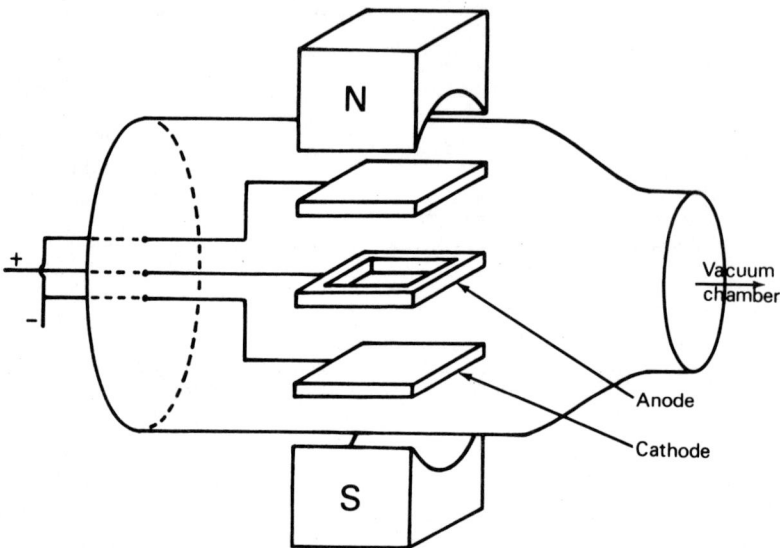

Figure 6.7.   Arrangement of a Penning ionisation gauge

to the vacuum chamber. A development of this basic arrangement is to use a flattened metal tube at earth potential as the cathode, with an anode ring mounted inside the tube. The cathode metal is often aluminium, nickel-plated copper, or stainless steel. A direct voltage of about 2 kV applied between the anode and cathode causes any free electrons in the region to move towards the anode. These will collide with gas molecules and produce ionisation. The heavy positive ions produced will move towards the cathode causing further ionisation on their way and, on collision with the cathode, will liberate more electrons, which in turn travel towards the anode and produce more ion pairs on their journey. Thus there is a positive ion current towards the cathode and a negative ion (electron) current towards the anode. If the anode and cathode are of different surface areas, then conduction will take place more easily in one direction than in the other. In the Penning gauge the anode has a relatively small surface area, which means that there is a much greater current density on its surface. If this current density exceeds a certain critical value on the small anode, but not on the larger cathode, it would require a higher voltage between the anode and cathode to maintain a negative ion current towards the anode than would be required for a positive ion current of the same magnitude towards the cathode. Consequently, there is a greater positive ion current than there is a negative one. The magnitude of this positive ion current is of course dependent on the number of ion pairs produced, which in turn is dependent on the number of molecules of gas present, that is, the pressure.

However, at low pressures the mean free path length is great, and therefore steps must be taken to ensure that collisions do occur. This is done by fitting a magnet with a flux density of about 5×10^{-2} tesla to the gauge, as shown in Figure 6.7. Thus electrons moving towards the anode travel in spiral paths and many pass right through the ring anode to be repelled again by the cathode. This may happen many times, the electrons spiralling back and forth through the anode before being finally captured, and hence travelling paths that may be several hundred times the direct path between a cathode and the anode. Again, as a gauge, it must be calibrated against a McLeod gauge and, since its lower limit is below that of the McLeod, some extrapolation is required. For pressures below about $10^{-5} \, N \, m^{-2}$, the relatively small number of gas molecules allows few collisions and the ion current becomes too small for practical measurement but, above this pressure, owing to the ruggedness of the construction and the lack of any hot filament to burn out, the gauge is found to be extremely useful and popular. There is, however, an upper limit to its pressure range of about $3 \, N \, m^{-2}$. This is due to the breakdown of carbon compounds, for example from pump oils, which causes contamination of the electrode surfaces, although the gauges are usually demountable and easily cleaned. To avoid contamination and the possibility of arcing, the gauge should not be left on when at atmospheric pressure.

A disadvantage of the Penning gauge is that it also acts as a pump and must therefore have a very high conductance connection to the vacuum chamber whose pressure is required, otherwise an anomalously low pressure will be registered. This pumping action arises from the bombardment of the cathode by heavy positive ions, causing metal atoms to be knocked, or sputtered, from the cathode surface. These metal atoms are deposited over the gauge walls and, by chemical reaction with the gas molecules, effectively remove them from the free volume and hence reduce the pressure.

A gauge working on a similar principle but having the advantage of greater accuracy is the *hot-cathode ionisation gauge*. Its construction is that of a conventional triode valve, as shown in Figure 6.8, where a central tungsten filament or emitter is surrounded by a grid of

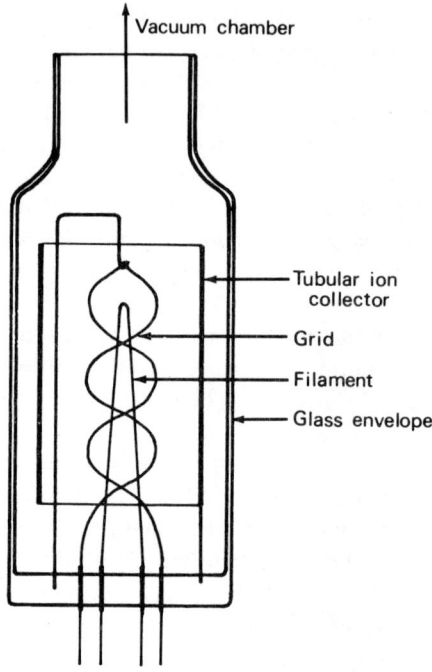

Figure 6.8.   Section through a hot-cathode ionisation gauge

molybdenum wire. Around this is a nickel cylinder which acts as an ion collector. A positive voltage of about 150 V is applied to the grid, causing electrons emitted by the hot filament to be accelerated towards the grid and to produce ionisation by collision of the gas molecules in their passage. Owing to the open structure of the grid, some of the electrons will pass through the grid to cause ionisation beyond, oscillating back and forth before final capture by the grid. The positive ions produced by ionisation of gas molecules go to the collector, which carries a voltage of about 20–30 V negative with respect to the filament. The constancy of electron production by the filament is maintained by adjusting the filament temperature by varying the voltage to maintain a constant emission current between the filament and the grid of about 1–5 mA. The pressure, which is a function of the number of positive ions produced, is determined from the ion current to the collector. It is usual for the emission current to be maintained constant electronically and the ion current to be passed to a d.c. amplifier before being registered by a meter, rather than simply using a microammeter.

An advantage over a cold-cathode ionisation gauge, apart from the greater accuracy, is the ease of outgassing. Molecules of gas tend to stick to cold surfaces and come off slowly with time; consequently, they give the same effect as a small leak. By heating the surfaces, these gas molecules can be driven off, or outgassed, and pumped away to cause no further trouble. To outgas a cold-cathode ionisation gauge a surrounding oven is required, whereas a hot-cathode ionisation gauge can be outgassed by passing an increased current through the filament to heat the gauge. Again, as with the cold-cathode ionisation gauge, a high conductance connection is required to the chamber of which the gas pressure is required as the gauge shows some pumping action due to gettering, that is, the using-up of free gas molecules by chemical reaction with metal surfaces and atoms. For this reason a nude gauge, that is, one without an envelope, is often used and mounted inside the actual vacuum chamber.

The upper limit of the gauge is about 1 N m^{-2}. In the region of this pressure, physical deterioration of the filament is rapid and it is soon likely to burn out. Thus the gauge must never be switched on when at atmospheric pressure. Owing to the usual electronic amplification of the ion current, the copious supply of electrons produced by the hot filament, and the open structure of the grid, a magnet to increase the path length of the electrons before capture is not generally required.

The lower pressure limit is set by the electrons striking the grid and causing it to emit soft x-rays. These x-rays in turn irradiate the ion collector, causing it to emit secondary electrons, so the positive ion current as recorded by the gauge appears greater by the amount of this electron current from the collector. At pressures in the region of 10^{-6} N m^{-2}, this residual electron current is comparable to the pressure-related positive ion current and at lower pressures becomes the overriding factor. Thus to extend the lower limit it is necessary to reduce this x-ray effect.

213

In the *Bayard–Alpert gauge* the configuration of the hot-cathode ionisation gauge is modified by reducing the ion collector in size from a cylinder to a wire, as shown in Figure 6.9, so as to reduce the area susceptible to x-radiation and so reduce the secondary electron emission from the collector. With this arrangement the lower limit of pressure

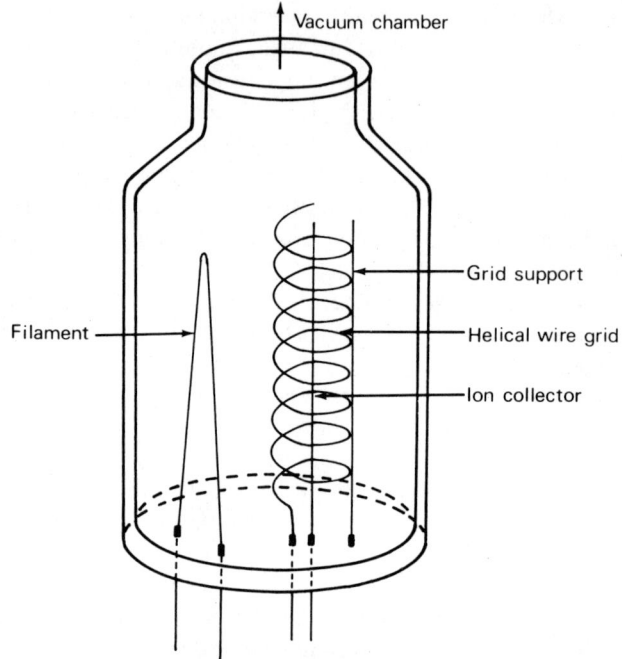

*Figure 6.9. Bayard–Alpert arrangement of a hot-cathode
ionisation gauge*

measurement set by x-ray emission is reduced to about 3×10^{-8} $N m^{-2}$. The upper pressure limit of about $1 N m^{-2}$ remains unchanged with the new configuration and is still set mainly by the risk of filament burn-out. An extension of about another order lower on the lowest pressure measurable can be achieved by using another single-wire electrode close to the grid and parallel to the ion collector. By varying the potential on this electrode between zero, with respect to the filament, and the potential of the grid, the positive ion current to the ion collector is modulated and can be selectively discriminated against with respect to the constant residual current due to x-ray emission.

214

As with the Pirani gauge, an ionisation gauge has different sensitivities for different gases, owing this time to differences in their ionisation characteristics. Usually the gauge is calibrated with nitrogen against a McLeod gauge so that, if it is to be used with some other gas, the recorded pressure must be multiplied by a suitable calibration factor.

6.5 Pumping Systems

To pump down a chamber from atmospheric pressure to the required degree of vacuum generally requires different types of pumps to cover different pressure ranges. To obtain a high vacuum it is necessary to use first some type of roughing pump, generally a mechanical one, to bring the pressure down to about 1 N m^{-2}, which brings the pressure into the range capable of being handled by the next stage pump, generally a diffusion pump. This type operates over the pressure range from about 1 N m^{-2} down to about 10^{-4} N m^{-2}. When it comes into operation, the system acts in a sense as a two-stage one, with the diffusion pump taking gas at low pressure from the chamber being evacuated and compressing it to a pressure that can be pumped by the mechanical pump. The mechanical pump is then said to 'back' the diffusion pump. Because a diffusion pump contains a hot working fluid, usually a silicone oil, it must not be exposed to air at atmospheric pressure. Thus a system of taps, or valves, is required as shown in Figure 6.10. Starting with all valves closed, the mechanical pump and Pirani gauge are switched on. The backing valve is then opened and the diffusion pump evacuated to a pressure of about 1 N m^{-2}. The diffusion pump heater may then be switched on, as well as the water cooling required at the top end of the pump. Now the backing valve may be closed, the roughing valve opened, and the chamber evacuated to a pressure of about 1 N m^{-2}.

Meanwhile, the diffusion pump will have heated up and become operational, so the roughing valve may be closed, the backing valve opened again, the baffle valve opened, and the ionisation gauge switched on. Thus the diffusion pump is now pumping the chamber by moving gas from the chamber to the bottom of the diffusion pump, where it is at sufficient pressure to be removed by the mechanical backing pump. Provided it is remembered that the hot diffusion pump must not be exposed to the atmosphere, the operation of the valves follow in a logical manner.

215

To let air into the chamber again, it is necessary only to close the baffle valve, switch off the ionisation gauge, and admit air to the chamber. Pumping-down again involves closing the air admittance valve and the backing valve, and roughing-down again. To close down the

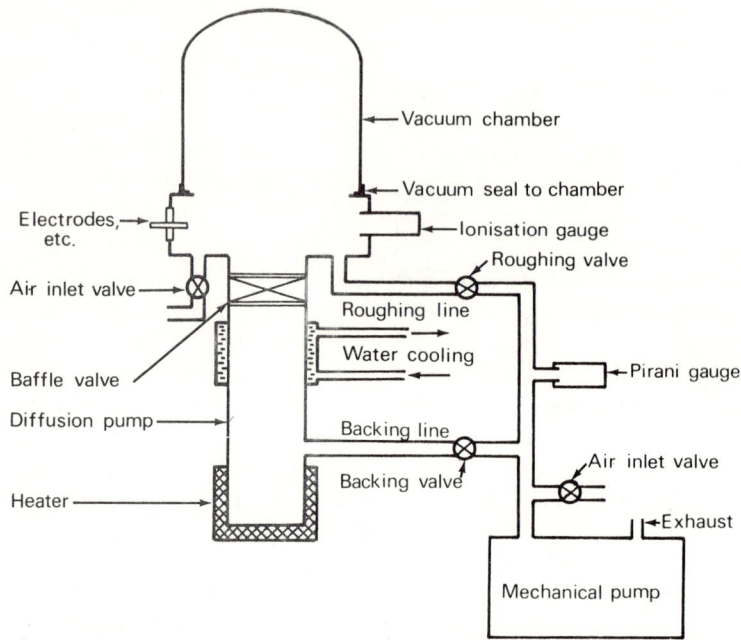

Figure 6.10. Arrangement of pumps and valves for a high vacuum pumping system

system finally, the diffusion pump should be allowed to cool down whilst being backed by the mechanical pump. When cold it should be isolated by closing the backing valve and baffle valve, if not already closed. On turning off the mechanical pump, the air admittance valve should be opened immediately, otherwise oil may be sucked up from the pump and flood the system.

When an ultrahigh vacuum is required with pressures down to about 10^{-8} N m^{-2}, then a different system of pumps is required. Although some improvement of the lower limit of a diffusion pump can be achieved with rather elaborate cold trapping techniques, the result is still not very satisfactory. Since presumably an ultrahigh vacuum is required to avoid contamination of surfaces, it is inherently better to

216

use an all-clean system where the pumps used do not require oil or mercury for their operation.

A typical arrangement for an ultrahigh vacuum pumping system is shown in Figure 6.11. A sorption pump is used to reduce the pressure of the complete system to about 10^{-1} N m^{-2} and the isolation valve is closed. The operating range of the ion pump is from about 10^{-1} N m^{-2} down to about 10^{-8} N m^{-2} although this can be improved upon with

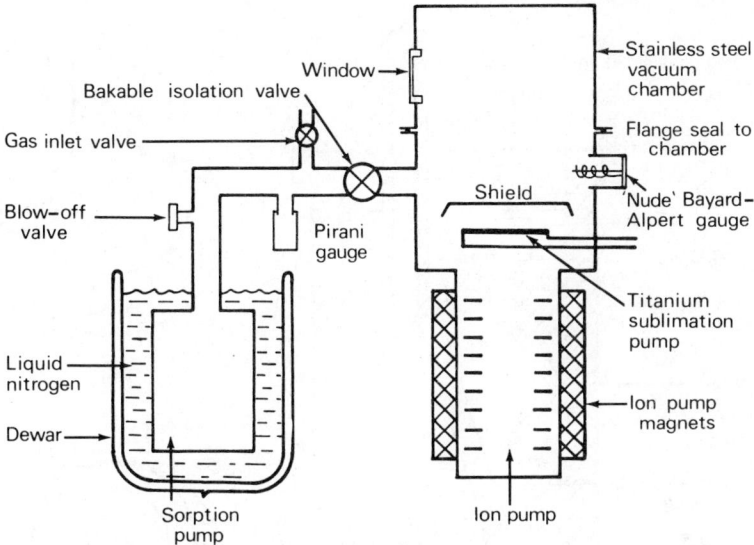

Figure 6.11. Arrangement of an ultrahigh vacuum system

special techniques and system design. At the starting pressure of about 10^{-1} N m^{-2}, the ion pump may not readily start to operate and therefore a titanium sublimation pump may be used to bridge the pressure gap until the ion pump comes into full operation. The sublimation pump is also useful in increasing the pumping speed at lower pressures.

In the next section we shall describe the working of the individual pumps mentioned and afterwards describe some of the vacuum accessories, such as valves and seals.

6.6 Vacuum Pumps

Let us describe first the pumps used in a conventional high vacuum system, that is, mechanical and diffusion pumps, and then the pumps used in a conventional ultrahigh clean vacuum system, that is, sorption, sublimation, and ion pumps.

Mechanical pumps are of many designs, depending on different manufacturers and the speed required. A common type is the *vane-type rotary pump*, shown diagrammatically in Figure 6.12. A rotor carrying a pair of spring-loaded sliding vanes is driven round inside a stator. Gas

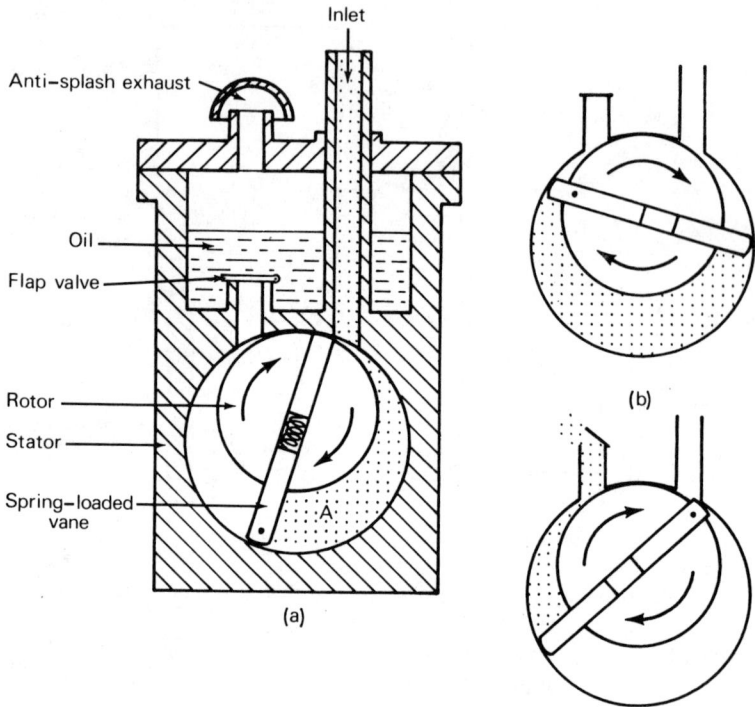

Figure 6.12. (a) Section through a vane-type rotary pump, and (b) subsequent positions of the rotor showing the gas being isolated and compressed

entering the pump will first occupy a volume marked A in the figure and will then be isolated by the vanes as the rotor rotates [Figure 6.12(b)]. Further rotation compresses this gas into a smaller volume until its pressure is above atmospheric and is forced out, escaping by lifting the flap valve. The pump is filled with oil to assist the sealing at the flap valve and at the sliding surfaces between the rotor, the vanes, and the stator. The lowest pressure attainable is limited by the back leakage of gas across the sliding oil seal between the rotor and stator, and is about 5×10^{-1} N m^{-2}. By coupling the outlet of one pump directly to the inlet of a second pump, a two-stage pump is formed. In this the pressure differential across the rotor–stator seal can be considerably reduced and with it, back leakage of gas. This enables a lower ultimate pressure of about 10^{-2} N m^{-2} to be reached.

The pumping capacity of such a pump can be specified in two ways. In Figure 6.12(a), the rotating vane is about to isolate the gas in the volume marked A. This volume of gas will then be compressed, to be finally ejected. Since there are two vanes, this cycle will occur twice for each revolution of the rotor. Thus, if V is the maximum volume of gas taken in, that is, the volume at A in Figure 6.12(a), the *displacement* of the pump is given by

$$\text{displacement} = 2nV \, \text{m}^3 \, \text{s}^{-1} \qquad (6.52)$$

where n is the number of revolutions of the rotor per second, and the volume V is measured in cubic metres. These pumps are manufactured with displacements up to about 0·2 m^3 s^{-1}, with a displacement of perhaps 1×10^{-3} m^3 s^{-1} being a more common laboratory size. Obviously the displacement is equal to the speed of the pump at atmospheric pressure, since it is the volume of gas removed by the pump in unit time when the gas at the inlet is at atmospheric pressure.

The other way of specifying the capacity is to quote the actual pumping speed with pressure, which for this type of pump is not constant. As the pressure at the inlet is reduced by the operation of the pump, the speed decreases. This is because of back leakage, outgassing of the oil, and the fact that the volume swept out by the vanes is not decreased to zero — there is still a finite volume leading up to the exhaust flap valve.

If the pump is used with an atmosphere which includes, for example, water vapour, then difficulties are encountered. On compressing the water vapour and gas mixture during the exhaust cycle of the pump, the saturation vapour pressure at the pump temperature may be exceeded and water will condense out. As free water it will be discharged to mix with the oil where, apart from causing serious deterioration of the oil, it will diffuse into the input side of the pump and return to vapour at the lower pressure. Thus it will cycle continuously between the exhaust and inlet sides of the pump and reduce the degree of vacuum obtainable, that is, the ultimate pressure will be increased. To deal with this problem of condensable vapours, *gas ballasting* may be employed. This adversely affects the compression ratio, that is, the ratio of the maximum to minimum volume swept out by the rotating vanes, but does allow the pumping of condensable vapours. The method is to allow dilution of the gas by admitting air through a non-return valve during the exhaust process. In Figure 6.13 the gas and vapour in the volume marked A are

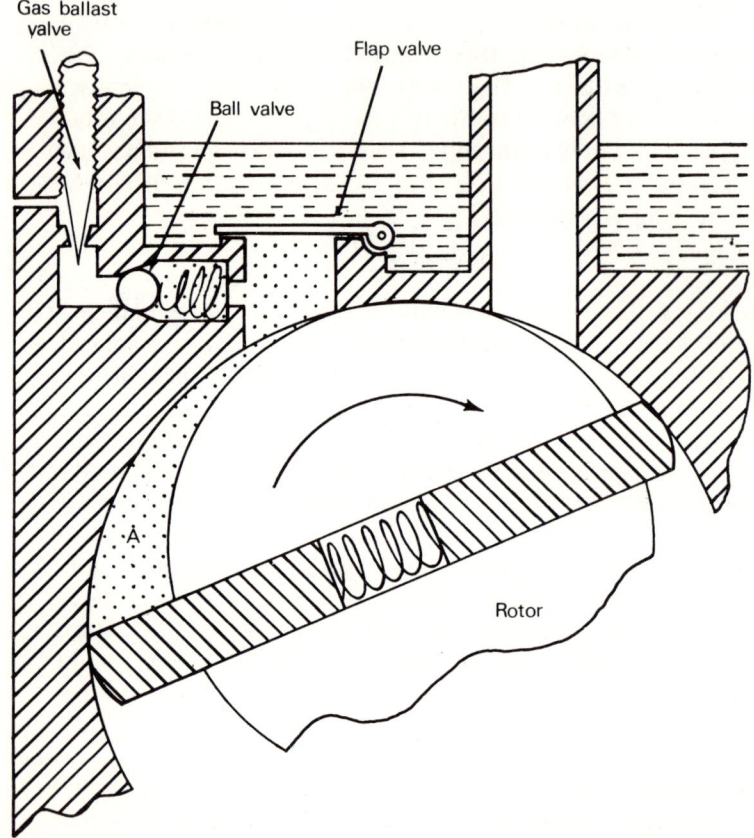

Figure 6.13. Part of section of vane-type rotary pump showing the modification of the exhaust to provide gas ballasting

at a pressure lower than atmospheric. Hence the ball valve will open and admit air at atmospheric pressure which dilutes the vapour and gas mixture. Further compression of this volume causes the ball valve to close and the exhaust flap valve to open. That is, the flap valve is opened and exhaust begins before the partial pressure of the water

220

vapour reaches the saturation value. Since the pumping efficiency is impaired by the gas ballasting, the gas ballast flow is adjustable to a maximum of about 10% of the pump displacement. As the vapour is cleared from the system, the gas ballast valve can be progressively closed to enable the ultimate vacuum pressure of the pump to be reached.

Taking over from the lower limit of the rotary pumps, the diffusion pump can take the pressure from about 1 N m^{-2} down to about 10^{-4} N m^{-2}, although by careful design and efficient cold trapping with liquid nitrogen several orders lower pressure can be obtained. Generally, however, this is not warranted as ion pumps can be used with much greater efficiency at the lower pressures. An arrangement of a simple diffusion pump is shown diagrammatically in Figure 6.14. An electric

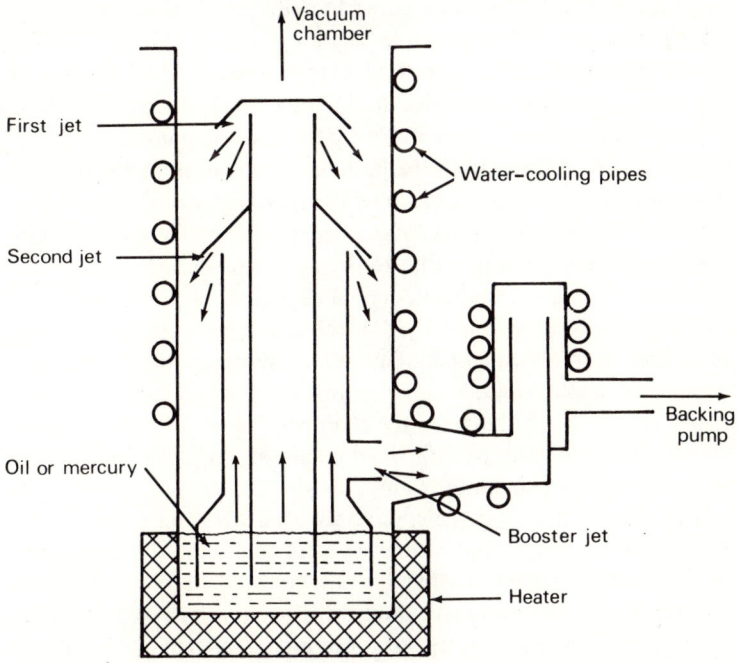

Figure 6.14. Operation of a diffusion pump shown diagrammatically

heater causes the pump fluid, generally a silicone oil although mercury is still used for special applications, to boil and produce a copious supply of vapour. This vapour is ejected at high speeds from the jets. Molecules of gas are caught up in the vapour jet stream from the first and second jets shown in the figure and carried towards the bottom of the pump by collisions with the fast-moving heavy vapour molecules. Vapour contacting the cold walls is condensed and runs down to the boiler for recycling. The action of the jets causes gas molecules to accumulate at the bottom of the pump, with a consequent increase in pressure so, for example, for an inlet pressure of 10^{-4} N m^{-2}, the pressure of the gas at the bottom of the diffusion pump may be raised to perhaps 1 N m^{-2} — a compression ratio of 10^4. At this higher pressure the gas can be effectively removed by a rotary pump. The connection to this backing pump usually involves some form of cooled labyrinth to condense oil vapour from the diffusion pump and return it, otherwise it would be pumped away by the backing pump. Also commonly incorporated is a booster jet, as shown in the figure, to assist the backing pump and enable the diffusion pump to continue to work against high backing pressures. If, however, the backing pressure is allowed to rise above a certain value, the *critical backing pressure*, the vapour jet will be unable to carry gas molecules up the increased pressure gradient and the pump will cease to function. Hence the backing pump must be capable of removing gas from the base of the diffusion pump at a sufficient rate to maintain a pressure less than the critical value.

If some of the vapour molecules in the jet stream leave the jet with a velocity in a direction opposite to the main stream, then *back streaming* occurs. The vapour molecules could also obtain this back velocity, perhaps by collision with the walls of the pump, but either way there will be migration of some of the pump fluid molecules back into the chamber being evacuated. To prevent this a *baffle* is required to prevent an optical line-of-sight between the chamber and the diffusion pump. This may be a labyrinth path or more commonly is the flap valve that isolates the chamber from the pump.

Although silicone oil is the commonly used diffusion pump fluid, mercury still has many applications. A diffusion pump should not be exposed to the atmosphere while hot, otherwise the pump fluid will be damaged. This is especially so for oil, which will oxidise and decompose to form a black sludge of little use for pumping. Mercury too will oxidise if exposed while hot but, since it is constantly being redistilled, it will recover its pumping efficiency provided it is not subjected to long and repeated exposure and, of course, it cannot decompose. The disadvantage of mercury is its high vapour pressure, so for pressures lower than about 10^{-1} N m^{-2} a liquid-nitrogen-cooled trap is required above the diffusion pump to condense the mercury vapour, whereas this is not necessary with silicone-oil-filled diffusion pumps. Simple water cooling is then quite sufficient as silicone oils are now manufactured with vapour pressures to cover all requirements with the choice being limited only by cost. Oils may also have a disadvantage when used with vacuum systems incorporating high voltages and electrical discharges. Under these conditions the oil is liable to be broken down into constituent atoms. The older hydrocarbon vacuum oils formed carbon deposits, whereas the normally used silicone oils form insulating films of silicone or silicon oxide, which may be as troublesome as the conducting carbon deposits. Thus mercury has an advantage in these circumstances and similarly for use with, for example, gas analysis apparatus since it does not decompose. Mercury in its turn has disadvantages when used with systems involving evaporated aluminium films as, unless elaborate cold trapping is used, the mercury vapour will destroy the newly formed aluminium film before it has had a chance to form a protective oxide layer.

As with rotary pumps, there are many variations in the design of diffusion pumps, which may have two, three, or four stages depending on the ultimate pressure and speed required and the backing pressure available. Jet designs may be modified to suit the different pump fluids and give greater pumping speeds over particular required pressure ranges.

The two pumps so far described can in combination give high vacuum conditions with pressures easily obtainable down to about 10^{-4} N m^{-2}, and with difficulties down to a couple of orders lower. They do, however, require a pumping fluid, whether it be oil or mercury, with its consequent vapour pressure limitations and the risk of contamination of clean surfaces. Although there are mechanical pumps which are clean, for example the *turbo-molecular pump* as designed by Becker, which works on the turbine principle and can attain pressures as low as 5×10^{-8} N m^{-2}, they generally have the disadvantages of being large and requiring very high rotational speeds with their inherent mechanical difficulties. Thus for preference a much simpler type of pump should be used when a clean system is required. The first in the series of commonly used clean pumps, which lead ultimately to ultrahigh vacuum conditions, is the *sorption pump*.

Sorption pumps can reduce the pressure in a system from atmospheric to about 10^{-1} N m^{-2}, depending on the particular design and operating conditions. Apart from being clean, containing no oil, they have no moving parts and are silent. The pump is simply a cylindrical container, usually in stainless steel, containing what is known as a *molecular sieve* material. This is commonly calcium alumino-silicate in the form of pellets about 3 mm in diameter. During the manufacture, the molecular sieve material is heated to remove the water of hydration, leaving a very porous structure with uniform pores about 5×10^{-10} m in diameter. When the material is cooled to liquid-nitrogen temperatures, gas molecules are adsorbed into the pellets and the pressure is rapidly reduced. The molecular sieve material is a poor thermal conductor and therefore the pump should be in the form of concentric cylinders to expose a large area to the liquid nitrogen, as shown in Figure 6.15.

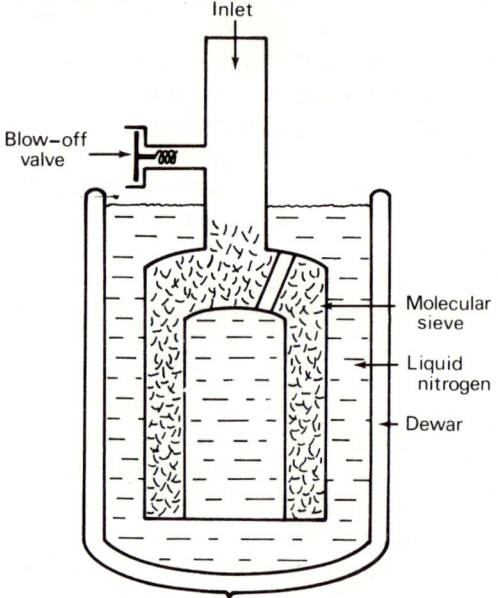

Figure 6.15. Section through a sorption pump

Liquid nitrogen is to be preferred to liquid air as the coolant, as with liquid air the nitrogen will boil-off first, leaving a strong concentration of liquid oxygen with its inherent danger of fire or explosion.

To reactivate the pump after use, it must be allowed to warm up. As this happens, the gas molecules are released from the molecular sieve pellets and the pressure rises above atmospheric, to be released by a blow-off valve. One disadvantage of this type of pump is that it can only adsorb a given quantity of gas during one cycle and is therefore unsuitable for pumping against a leak. Sorption pumps also tend to sorb preferentially water vapour and to desorb it less readily at room temperature. Consequently, it is occasionally necessary to reactivate completely the molecular sieve material by heating the pump at atmospheric pressure to about 300°C for a couple of hours. Providing these pumps are treated reasonably, for example the pellets not being contaminated with oil, and are completely reactivated very occasionally as needs demand, then their life is virtually unlimited.

The pressure in the vacuum chamber having been reduced to about 10^{-1} N m^{-2}, it is almost down to the firing pressure of the ion pump. To help bridge the small pressure gap and get the ion pump quickly to its more efficient operating pressures, a *titanium sublimation pump* can be used. Before using this, however, it is necessary to close the valve isolating the sorption pumps from the vacuum chamber, otherwise the gas adsorbed by the molecular sieve material will be pumped back into the chamber.

The sublimation pump, or *getter pump* as it is sometimes called, works on the simple principle of evaporating titanium from a hot source and letting it condense onto a cold surface, for example the walls of the chamber or a conveniently placed shield. The freshly deposited clean titanium atoms then pump by a gettering action, removing gas molecules from the free volume by chemically reacting with them. This gettering action will thus depend on the number and type of molecules and their sticking coefficients so, for example, the pumping speed is fast for molecules such as CO, CO_2, and H_2, but virtually zero for the inert gases such as argon. Although designs of pumps are available for continuously feeding titanium wire onto a hot anvil, they tend on the whole to be troublesome to operate. A simpler system is to use a single tungsten wire, about 10 cm long, coated with titanium. By passing a heavy current through the wire, the titanium is heated and caused to evaporate. Usually several of these titanium—tungsten wire elements are mounted side-by-side so that, when the titanium on one element is used up, the heating current can be switched to the next element.

Once the ion pump starts to operate, at a pressure perhaps a little below 10^{-1} N m^{-2}, the sublimation pump is required as only an occasional boost to the ion pump, which will take the pressure down to about 10^{-8} N m^{-2}, with a maximum pumping speed occurring over the range of about $10^{-3}-10^{-6}$ N m^{-2}. This lower limit of 10^{-8} Nm^{-2} is mainly set by the cleanliness of the system and the slow outgassing

of the surfaces. The principle of operation of the *ion pump* is the same
as that of the cold-cathode ionisation gauge. Instead of the one cell
anode between two plate cathodes of the ion gauge, many cells are
arranged together in the pump, as shown in Figure 6.16. In the ion

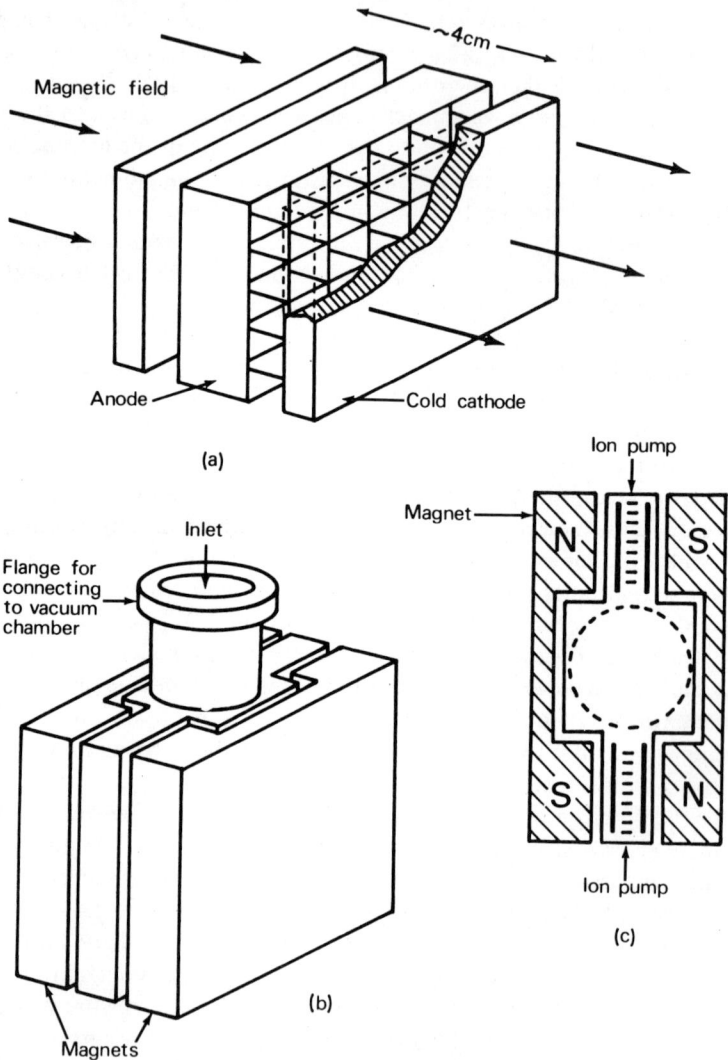

*Figure 6.16. (a) The arrangement of a multi-cell cold-cathode ion pump,
(b) its practical arrangement, which is shown in section in (c)*

pump, or *Penning pump* as it is also called, the anode is usually of titanium but may be of some non-magnetic alloy, while the flat plate cathodes are of titanium, 1—3 mm thick. A voltage of up to about 10 kV is then applied between the anode and cathodes, and a permanent magnet is fixed to the outside of the pump casing to provide a magnetic flux density of about 0·2 tesla. This, as with the Penning gauge, is to provide a long spiral flight path for the ions and thus increase the chances of collisions with gas molecules to produce further ionisation. The pump casing is of stainless steel, with the smaller pumps being sometimes of glass.

The magnets, being mounted externally, are removable, which may be of assistance in installing the pump as the large ones can be quite heavy. Occasionally the pump may require baking to about 500°C for cleaning purposes, when the magnets will have to be removed as they may otherwise be damaged.

As with the Penning gauge, a discharge will be set up with the electrons moving towards the anode and the positive gas ions towards the cathodes. At the cathodes the arriving heavy positive ions will produce various effects. They will cause titanium atoms to be knocked off the cathode and ejected, a process termed *sputtering*; they will cause gas molecules adsorbed on the surface of the cathode to be ejected, that is, gas sputtering; and finally, they will cause electrons to be ejected from the cathode. These ejected electrons will then proceed on spiral paths towards the anode and will provide more ions by collision with gas molecules.

The main pumping action is due to the gettering action of the sputtered titanium which is deposited mainly on the large surface area of the anode. Here it pumps by forming a titanium compound with the gas molecules that collide with it. The gas molecules are thus removed from the active volume and consequently the pressure is reduced. It has been estimated that from half to one molecule of gas is removed from the active volume for each atom of titanium sputtered. The inert gases, which of course do not form compounds with titanium, are pumped by being adsorbed onto the electrode surfaces and buried by the arriving titanium atoms.

An ion pump has a number of advantages over, for example, a diffusion pump. Perhaps the most important of these is the cleanliness due to the absence of any pumping fluid, which also means that no baffles or cold traps are required. Since the principle of operation is virtually that of a Penning cold-cathode ionisation gauge, the discharge current is a function of the pressure and, with suitable calibration, the pump can also be used as a pressure gauge although it may well record a slightly lower pressure than that existing in the chamber. To reach the ultimate pressures, it is necessary to drive off the adsorbed gas molecules from all the surfaces in the system, otherwise these will detach themselves over a long period and increase the pressure. To drive off these gas molecules, it is necessary to bake the complete vacuum system to about 300°C and hold it at this temperature for some hours. By allowing the pump to cool quicker than the rest of the system, gas molecules will tend to be driven towards the pump and thence removed. Baking is usually carried out with an oven assembly that can be lowered over the complete vacuum system or with strip heaters fixed to the system at strategic places.

The lifetime of the pump is mainly dependent on the lifetime of the titanium cathode plates. With operation at a pressure of about 10^{-4} N m^{-2} the lifetime may be perhaps 50 000 h. At higher pressures the lifetime may be drastically reduced by the flaking-off of titanium from the anode, which may short out the electrodes, and by severe erosion of the cathodes. Consequently, they should not be used in systems where the pressure will normally be above about 10^{-3} N m^{-2} or where the pressure is being cycled repeatedly. They should also not be used with, for example, a rotary backing pump without some form of cold trapping, otherwise the hydrocarbon vapour will poison the electrodes and seriously affect their sputtering properties.

A disadvantage of ion pumps is the possibility of *argon instability*. This is due to inert gases, of which argon is the most common, being pumped mainly by adsorption on the electrode surfaces. Argon atoms adsorbed on the cathode will be released again by gas sputtering processes due to arriving ions of other gases. Unfortunately, these argon atoms tend to be released together in bursts causing pressure instability. This tendency may, however, be reduced by a slotted design of cathode plate or by a triode arrangement of electrodes.

6.7 Vacuum Components

The constructional material of the vacuum system is determined mainly by the ultimate vacuum pressure requirements. All-glass systems are used for specialised processes, which may even employ all-glass diffusion pumps, for example, but generally a metal system is used,

228

with the possibility of a glass bell-jar for the vacuum chamber. The most suitable metal for piping in a high vacuum system is copper. Copper piping is seamless, is available in a range of diameters, and is easily bent and soldered. Brass is often used for the base plate of the vacuum chamber but, owing to the zinc in brass, it is not suitable for use in systems where elevated temperatures may be encountered. Having been heated and lost zinc, the brass may well be porous and for this reason it is often nickel plated before use. For ultrahigh vacuum systems which require baking, possibly occasionally to 500°C, the constructional material must retain its mechanical strength while evacuated and, more important, must have a low vapour pressure. It must also be impervious to the diffusion of gases through it. For these reasons, stainless steel has become the most popular material, even though it is mechanically a relatively difficult material to work and must be argon arc welded to obtain good vacuum-tight joints.

Synthetic rubber, usually Neoprene or Viton, can be used for seals between demountable components and in valves not subject to elevated temperatures and where its vapour pressure is acceptable as in high vacuum systems. For ultrahigh vacuum systems, the sealing material must be able to withstand baking and must have a sufficiently low vapour pressure. This limits the sealing material to a metal, with copper or gold being the usual materials. The use of these materials for sealing is illustrated in Figure 6.17, which shows one type of O-ring seal for joining two pipes for high vacuum use and a knife edge and copper seal for ultrahigh vacuum use.

In addition an L-section gasket of synthetic rubber is used for sealing a bell-jar chamber to a base plate in a high vacuum system, and usually a gold wire seal for sealing the chamber cover in an ultrahigh vacuum system. A gold wire seal is simply a ring of annealed gold wire squashed between two flanges as the flanges are bolted together. When using O-ring seals, it is important that the correctly dimensioned groove is used to retain the ring. Sealing is here due to squashing the ring to about 80% of its original diameter and allowing the correct space in the groove for the displaced volume. A smear of silicone grease around the O-ring is usually used to assist the seal.

Little has been said concerning cleanliness in a vacuum system but obviously anything that has a higher vapour pressure than the ultimate pressure required in the system must be avoided. Thus, for example, greasy finger-prints must not be left in a system in which it is hoped to attain ultrahigh vacuum pressures. Likewise, roughly machined surfaces

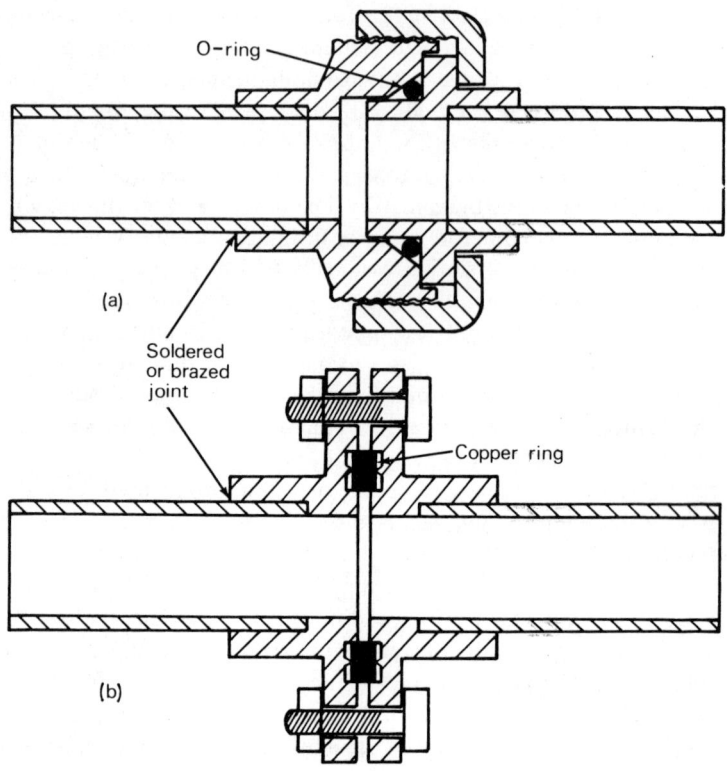

O-ring

(a)

Soldered
or brazed
joint

Copper ring

(b)

*Figure 6.17. Section through (a) a demountable O-ring tube coupling, and
(b) a knife edge and copper ring coupling for ultrahigh
vacuum use*

and blind screw holes must be avoided otherwise gas molecules swaged
into the surface or adsorbed will steadily leak into the active volume
to raise the pressure, as will the gas molecules trapped under the screws.
For the very lowest pressures, surfaces must be electrolytically cleaned
and the use of any hydrocarbon solvents avoided. For high vacuum
conditions such stringent care is not required, nevertheless care must be
taken if a sufficiently low pressure is to be achieved.

FURTHER READING

Chapters 1, 2, and 3

HALL, I. H., *Deformation of Solids*, Nelson, London (1968).
HAUSNER, H. H. (Ed.), *Modern Materials*, Vols 1–2, Academic Press, London (1960).
KELLEY, A., *Strong Solids*, Oxford University Press, London (1966).
POTMA, T., *Strain Gauges*, Iliffe, London (1967).
TIMOSHENKO, S., and GOODIER, J. N., *Theory of Elasticity*, McGraw-Hill, New York (1951).
TIMOSHENKO, S., and YOUNG, D. H., *Elements of Strength of Materials*, Van Nostrand, New Jersey (1962).
VAN VLACK, L. H., *Elements of Materials Science*, Addison-Wesley, Massachusetts (1964).

Chapters 4 and 5

DUNCAN, W. J., THOMAS, A. S., and YOUNG, A. D., *Mechanics of Fluids*, Arnold, London (1970).
EIRICH, F. R. (Ed.), *Rheology*, Vols 1–3, Academic Press, New York (1956).
MASSEY, B. S., *Mechanics of Fluids*, Van Nostrand, London (1968).
WHITAKER, S., *Introduction to Fluid Mechanics*, Prentice-Hall, New Jersey (1968).
WILKINSON, W. L., *Non-Newtonian Fluids*, Pergamon, London (1960).

Chapter 6

ROBERTS, R. W., and VANDERSLICE, T. A., *Ultrahigh Vacuum and its Applications*, Prentice-Hall, New Jersey (1964).
SPINKS, W. S., *Vacuum Technology*, Chapman & Hall, London (1963).
YARWOOD, J., *Highvacuum Technique*, Chapman & Hall, London (1961).

Appendix One

SI UNITS AND CONVERSION FACTORS

Basic SI Units

Quantity	Unit	Unit symbol
Length	metre	m
Mass	kilogram	kg
Time	second	s
Electric current	ampere	A
Absolute temperature	kelvin	K
Luminous intensity	candela	cd

Derived Units Having Special Names

Quantity	SI unit	Unit symbol
Force	newton	$N = kg\ m\ s^{-2}$
Work, energy, heat	joule	$J = N\ m$
Power	watt	$W = J\ s^{-1}$
Electric charge	coulomb	$C = A\ s$
Electric potential	volt	$V = W\ A^{-1}$
Electric capacitance	farad	$F = A\ s\ V^{-1}$
Electric resistance	ohm	$\Omega = V\ A^{-1}$
Frequency	hertz	$Hz = s^{-1}$
Magnetic flux	weber	$Wb = V\ s$
Inductance	henry	$H = V\ s\ A^{-1}$
Luminous flux	lumen	$lm = cd\ sr$
Illumination	lux	$lx = lm\ m^{-2}$

Conversion Factors

Length

$$1 \text{ Å} = 10^{-10} \text{ m}$$
$$1 \text{ in} = 0.0254 \text{ m}$$
$$1 \text{ ft} = 0.3048 \text{ m}$$
$$1 \text{ mile} = 1.6093 \text{ km}$$

Velocity

$$1 \text{ ft s}^{-1} = 0.3048 \text{ m s}^{-1}$$
$$1 \text{ mile h}^{-1} = 0.4470 \text{ m s}^{-1}$$

Area

$$1 \text{ in}^2 = 6.4516 \times 10^{-4} \text{ m}^2$$
$$1 \text{ ft}^2 = 0.0929 \text{ m}^2$$
$$1 \text{ mile}^2 = 2.5900 \times 10^6 \text{ m}^2$$

Volume

$$1 \text{ litre} = 10^{-3} \text{ m}^3$$
$$1 \text{ pint} = 5.6826 \times 10^{-4} \text{ m}^3$$
$$1 \text{ UK gal} = 4.5461 \times 10^{-3} \text{ m}^3$$
$$1 \text{ in}^3 = 1.6387 \times 10^{-5} \text{ m}^3$$
$$1 \text{ ft}^3 = 0.028 \, 32 \text{ m}^3$$

Mass

$$1 \text{ lb} = 0.453 \, 592 \text{ kg}$$
$$1 \text{ ton} = 1.016 \, 05 \times 10^3 \text{ kg}$$

Density

$$1 \text{ g cm}^{-3} = 10^3 \text{ kg m}^{-3}$$
$$1 \text{ lb in}^{-3} = 2.7680 \times 10^4 \text{ kg m}^{-3}$$
$$1 \text{ lb ft}^{-3} = 16.0185 \text{ kg m}^{-3}$$

Force

$$1 \text{ dyn} = 10^{-5} \text{ N}$$
$$1 \text{ kgf} = 9.8066 \text{ N}$$
$$1 \text{ pdl} = 0.1383 \text{ N}$$
$$1 \text{ lbf} = 4.4482 \text{ N}$$
$$1 \text{ tonf} = 9.9640 \times 10^3 \text{ N}$$

Pressure, stress

$$1 \text{ mmHg} = 133 \cdot 322 \text{ N m}^{-2}$$
$$1 \text{ torr} = 133 \cdot 322 \text{ N m}^{-2}$$
$$1 \text{ in H}_2\text{O} = 249 \cdot 1 \text{ N m}^{-2}$$
$$1 \text{ bar} = 10^5 \text{ N m}^{-2}$$
$$1 \text{ dyn cm}^{-2} = 10^{-1} \text{ N m}^{-2}$$
$$1 \text{ kgf cm}^{-2} = 0 \cdot 098 \ 07 \text{ N m}^{-2}$$
$$1 \text{ lbf in}^{-2} = 6 \cdot 8948 \times 10^3 \text{ N m}^{-2}$$
$$1 \text{ tonf in}^{-2} = 1 \cdot 5444 \times 10^7 \text{ N m}^{-2}$$

Viscosity

$$1 \text{ poise (g cm}^{-1} \text{ s}^{-1}) = 10^{-1} \text{ kg m}^{-1} \text{ s}^{-1}$$
$$= 10^{-1} \text{ N s m}^{-2}$$
$$1 \text{ stokes (cm}^2 \text{ s}^{-1}) = 10^{-4} \text{ m}^2 \text{ s}^{-1}$$

Appendix Two

USEFUL PHYSICAL CONSTANTS AND VALUES

Quantity	Symbol	Value
Atomic mass unit		$1·6604 \times 10^{-27}$ kg
Avogadro's constant	N_A	$6·0225 \times 10^{26}$ kg-mole^{-1}
Boltzmann's constant	k	$1·3805 \times 10^{-23}$ J K^{-1}
Electron charge	e	$1·6021 \times 10^{-19}$ C
Electron rest mass	m_e	$9·1091 \times 10^{-31}$ kg
Electron volt	eV	$1·6021 \times 10^{-19}$ J
Gravitational constant	G	$6·670 \times 10^{-11}$ N m^2 kg^{-2}
Planck's constant	h	$6·6256 \times 10^{-34}$ J s
Standard gravitational intensity	g	9.8067 m s^{-2}
Universal gas constant	R	$8·3143 \times 10^3$ J kg-mole^{-1} K^{-1}
Velocity of light *in vacuo*	c	$2·9979 \times 10^8$ m s^{-1}

Density of mercury at 0°C	$1·3595 \times 10^4$ kg m^{-3}
Standard atmospheric pressure	$1·0132 \times 10^5$ N m^{-2}
Vapour pressure at 20°C, mercury	$0·1601$ N m^{-2}
Vapour pressure at 20°C, water	$2·335 \times 10^3$ N m^{-2}
Velocity of sound at 20°C, air	$3·434 \times 10^2$ m s^{-1}
Viscosity at 20°C, air	$1·81 \times 10^{-5}$ N s m^{-2}
Viscosity at 20°C, water	$1·0019 \times 10^{-3}$ N s m^{-2}
Young's modulus E, copper	$1·298 \times 10^{11}$ N m^{-2}
Young's modulus E, steel	$2·1 \times 10^{11}$ N m^{-2}

INDEX

237